Field and Laboratory Exercises in Animal Behavior

Field and Laboratory Exercises in Animal Behavior

Chadwick V. Tillberg
Linfield College, McMinnville, Oregon, USA

Michael D. Breed
University of Colorado, Boulder, Colorado, USA

Sarah J. Hinners
University of Colorado, Boulder, Colorado, USA

AMSTERDAM • BOSTON • HEIDELBERG • LONDON • NEW YORK • OXFORD
PARIS • SAN DIEGO • SAN FRANCISCO • SINGAPORE • SYDNEY • TOKYO
Academic Press is an imprint of Elsevier

Academic Press is an imprint of Elsevier
84 Theobald's Road, London WC1X 8RR, UK
30 Corporate Drive, Suite 400, Burlington, MA 01803, USA
525 B Street, Suite 1900, San Diego, CA 92101-4495, USA

First published 2007

Notice
No responsibility is assumed by the publisher for any injury and/or damage to persons
or property as a matter of products liability, negligence or otherwise, or from any use
or operation of any methods, products, instructions or ideas contained in the material
herein. Because of rapid advances in the medical sciences, in particular, independent
verification of diagnoses and drug dosages should be made

British Library Cataloguing in Publication Data
A catalogue record for this book is available from the British Library

Library of Congress Cataloging-in-Publication Data
A catalog record for this book is available from the Library of Congress

ISBN: 978-0-12-372582-0

For information on all Academic Press publications
visit our web site at books.elsevier.com

Printed and bound by CPI Group (UK) Ltd, Croydon, CR0 4YY
Transferred to Digital Print 2011

Working together to grow
libraries in developing countries

www.elsevier.com | www.bookaid.org | www.sabre.org

ELSEVIER   BOOK AID
            International   Sabre Foundation

Contents

Contents

Preface

Actually seeing animals behave and being able to do your own studies of animal behavior is so much more exciting and interesting than listening to lectures. To really learn about animal behavior you must make your own observations and interpretations. Watching behavior shows you both general patterns and variation from the generalities. Forming your own hypotheses and making your own interpretations sharpens your scientific skills. In the laboratory part of this course, we try to give you the freedom to make your own observations and to generate hypotheses, within the framework of behavioral topics, such as pheromones, mate selection, and territoriality. Comparing your results with those of other students will give you an even better understanding of how variable behavior can be, and of how observers may come to different conclusions when they watch the same behavioral acts.

Our approach to teaching laboratory and field animal behavior has two major pieces. The first is a series of structured explorations employing diverse organisms – insects, fish, and birds – to develop a skills base that covers a wide variety of behaviors. The second provides the student with an opportunity to pursue an aspect of behavior in depth by doing his or her own research project. Together, these pieces, one that is well-structured and the other that requires you to work more independently, provide a solid foundation of practical experience with animal behavior.

In the laboratory part of the course, we try to give students the freedom to make their own observations and to generate and test hypotheses within the framework of behavioral topics. In addition to experiments and field studies, we explore behavioral theories through games, role-playing, and hypothetical scenarios. As students compare their own results with others', they gain a better understanding of how variable behavior can be, of how observers may come to different conclusions when they watch the same behavioral acts, and of how the same set of circumstances in nature may give rise to very different behavioral responses.

This manual contains more labs than most courses will be able to cover. The labs that the instructor does not choose to use can serve as valuable resources for techniques and ideas for individual projects and independent investigations. The labs often include hand-in worksheets, as well as some oral presentations and written laboratory reports. The worksheets give instructors the opportunity to collate data from all the students, yielding a larger and more easily-interpreted datasets than any individual student could generate. These datasets are the basis for post-lab discussion and interpretation. Concluding discussions are the key to the labs, as they allow students to fully integrate their knowledge of behavior with what they have been observing. Some of the worksheets also contain questions that build on these laboratory discussions.

Above all, the labs should be interesting and engaging. The study of behavior is one of the most fascinating areas in modern science, and we hope that the labs convey that fascination.

Section 1
Independent Research in Animal Behavior: Getting Started

Chapter 1
Independent research projects in animal behavior

1 Why do an independent research project?

At the college or university level, most learning takes place in a relatively passive manner. As a student, you take in information from lecture or through reading; even in labs the activities are quite structured. Designing and conducting your own research project is an opportunity to take your learning to a whole new level and develop the skills that come from truly practicing science. This is an excellent preparation for anyone considering graduate school or a career in biological research. In addition, it's an opportunity to obtain hands-on experience by observing and analyzing an aspect of behavior that interests you; it's often a chance to work outdoors, and doing a project will also help you to develop your communication skills when you present your results to your peers.

Doing an independent research project is challenging. It requires creative thinking, attention to detail, and resolve. Everyone will encounter difficulties somewhere along the way, and the ability to adapt and solve problems is the key to success. Careful planning, familiarity with the primary literature, and explicit hypotheses will help get you off to a good start. It is important to communicate with your instructor at every step along the way to make sure you don't get bogged down or go off track.

II The stages of a research project

In this section we go through the process of developing a project, starting with a quick overview of the scientific method, and then progressing to more detailed advice on how to get started. Your goal is to choose a topic for your project, but first you need an overview of the scientific method. For many students this will be a review, but in our experience many of you will have been taught in a way that understates the importance of observation as the beginning maneuver in the scientific process.

A Scientific method and hypothesis testing: an overview

The scientific method is easily described as a stepwise process. You start with an observation and turn that observation into a question. The question is then recast as a hypothesis or set of hypotheses, which are more formal, testable expressions of the question. You then design a program of data collection or experiments to test your hypotheses. Finally, once you have the data, you analyze and interpret your results.

For example, you may have observed a squirrel burying acorns. You wonder why it would bury, rather than eat, the acorn; this is your question. You form the hypothesis that the squirrel buries the acorns so they can be collected again at a later date, when the squirrel is hungry. This leads to a plan of observation to determine, if, in fact, the squirrel comes back later and unburies the acorns. If it does, then you conclude that the squirrel has buried the food for later use. This description of the scientific process is deceptively simple; books by Carey (2003) and Gauch (2003) are very useful resources for more detailed discussions.

The challenge, of course, is finding a topic; the next section will give you some guidance on this difficult process.

B Observation: the first step in choosing a project topic

It should be obvious at this point that you need to make observations to start yourself towards a project. Observations can come from a variety

Box 1.1 A quick exercise in scientific observation.

Take a notepad, something to write with, and yourself on a walk. Whenever you see a non-human animal, stop and write down questions about what the animal is doing. Watch its behavior closely. Does it do anything puzzling or does it behave in ways that surprise you? A good goal is to observe four to six different species, and to ask five questions about each. Even in a very urban environment you should be able to easily find animals like dogs, pigeons, starlings, squirrels, and a variety of insects. You can do this on your own, or as part of a class exercise. If your class does this, you may want to compile the questions from the class and choose a few to discuss how you would transform the question into a hypothesis and then into an experiment.

of sources: (1) you can go into the field and make your own observations (see Box 1.1), (2) you can read published descriptions of behavior and use these as surrogates for making your own observations, or (3) your instructor can suggest projects, based on his or her observations. We prefer the first option, as you'll learn more making your own observations, but time or other constraints may cause you to choose the second or third routes to observation.

C Turning observations into questions

Regardless of how you obtain your initial observations, at some point you'll need to focus in on a question, and you'll find that there are two angles from which to approach choosing a topic. First, you may have a question or concept that has fascinated you, but you don't have an animal chosen yet for asking the question. For example, you may know that you would like to study foraging behavior, but you don't know what animal to use. Alternatively, you may have chosen an animal that grips you, but you don't know exactly what topic concerning that animal's behavior will make a good project. For example, you may have easy access to a group of animals, and this accessibility may make you decide to work on that species. In the end, you need both a question and a system, such as "How does the presence of predatory birds affect foraging behavior in checkerspot butterflies."

D What if you have a topic, and need a study animal?

If you've first chosen a topic, then you need to find an animal in which to explore that topic (Box 1.2). The most common sources for class projects include pets, abundant and easily-accessible wildlife, or lab organisms, which can be obtained from scientific supply companies. In any case, you'll have to do some library or web-based research to find out which organism would be appropriate to test your question. In many cases an attractive or abundant species turns out to be not at all suitable for the study you have in mind, and you need to avoid trying to bend a species' biology to your experimental goals.

As you choose an animal, there are some important considerations about the ethical treatment of animals to keep in mind. The well-being of your test subjects is one of the most important considerations in doing animal behavior research. Ultimately you need to follow your own moral compass in terms of appropriate use of animals in experiments, but you also must conform to any school, local, state, or federal regulations that might govern your experiment. You should consult with your instructor about these regulations and how they might affect your plan. An important aspect of studying behavior is that healthy test subjects yield more reliable data, so the success of your project may hinge on the health of the animals in your study.

For good reasons, your educational institution probably has its own set of guidelines for using animal subjects in research, and if you plan to use lab animals on campus you will probably have to do some paperwork and obtain approval. Similarly, most public lands are subject to legislation that limits the extent to which you may interact with wild animals. Many regulations and laws extend to the treatment of animals on private property; don't assume that being on private property exempts you from obtaining permits and treating animals appropriately.

Box 1.2 Sources of subjects.

- Pets
- Wildlife (this includes the less "wild" forms of wildlife such as pigeons or squirrels)
- Zoos, public aquaria, public insectaria, animal shelters or rescue operations, farms, and so on
- Biological supply houses such as Carolina or Ward's

In general, animal behaviorists must weigh the importance of the knowledge to be obtained from their research against whatever suffering the animals may endure, and they must take every possible precaution to minimize that suffering. Considering the scope of this project, and for the sake of simplicity, let us simply agree upon the following guidelines:

- Animals should not be subjected to pain as part of the experiment.
- Animals should not be subjected to deprivation (sleep, nourishment, etc.).
- Social animals should not be subjected to prolonged periods of isolation.
- Wild animals should be observed from a distance. (In general, trapping or marking wild mammals or birds requires permits from local and federal agencies and is outside the scope of your project.)
- Animals should not be placed in situations where one individual causes physical harm to another.

For more information on the ethics of animal behavior studies, the Animal Behavior Society has guidelines which are available at: http://www.animalbehavior.org/ABS/Handbook/abspolicy99.html#treatment

Another important issue is obtaining the appropriate permissions to do your study. Your instructor should be able to guide you in this. Students doing projects should take care to conform with federal, state, and local regulations governing animal welfare. Your instructor can help you to understand the regulations and doing the paperwork. On many campuses, students in biology courses may use campus property for projects, as long as no property damage results. Activities on private lands always require the permission of the landowner, and working on private land does not free you from local, state, and federal environmental regulations.

Handling of "lower" or non-endothermic vertebrates is less strictly regulated, but you should still treat any fish, amphibia, or reptiles that you might employ in your project with respect. Use of invertebrates (e.g. insects) in projects is fine; but again, the animals should be treated with respect.

You also need to carefully consider sample size, including the number of animals you'll need for your study. There is often an incredible amount of behavioral variation among individual organisms within a species. Researchers deal with this variation by testing many individuals and averaging the responses. When designing your study, it is important that you have enough subjects to properly test your hypothesis. Conducting your study using only your dog as a subject will not produce meaningful results; it will simply tell you about your dog. It is difficult to say how many subjects you will need, and in the end it will be a balance between the need for scientific rigor and the practical realities of how many animals you have access to. Also, you should keep in mind that if you are using two different treatments, you will need subjects for each treatment, thus doubling the number of animals required. In general, for small studies, many scientists apply a simple "rule of 10", or 10 subjects per treatment. You should discuss your plans with your instructor to make sure that you have an adequate sample size.

The major pitfall for you to avoid in choosing a project is attempting to study animals that are rare or difficult to observe. Bobcats, for example, have fascinating behavior but are so secretive that finding subjects for study would be difficult, at best. Most hypotheses concerning behavior can be tested in a variety of common animals, and your project will be more rewarding if you don't create barriers for yourself in your choice of study species.

E What if you have study animal but no topic?

In the case when you have the organism but no topic, you have a couple of options for focusing in on a question. First, you can do some library research to find out what is behaviorally interesting about this species or breed. What have other people studied in this organism? Second, you can spend some time observing the organism and generate some questions on your own. For example, if you want to study the common fox squirrel (abundant on most college campuses), you could spend some time sitting on the grass just seeing what they do. All of the considerations concerning the use of animals in projects discussed above apply, of course, to projects that are generated in this way.

If you have neither a study animal nor a topic, go directly to the next stage, there are several options. First and foremost, don't despair. Try spending some time with your textbook and doing some literature browsing in the library; *something* (a topic or animal) should catch your interest. Or, spend some time looking around, as suggested in Box 1.1, and consider what organisms are easily accessible or observable.

Background research – what do you know about this topic?

Now that you have a topic, you'll need to go to the library for a bit of background research. We advocate finding out enough about what's known concerning your study animal and topic to do our project, but not so much that you are overwhelmed by what is already known. Much good science results from work that is done from a naïve point of view;

if you bring a fresh, unbiased viewpoint to a project you may find out something that previous investigators missed or misinterpreted.

It's also important to be able to put your research into the larger context of what we already know about this topic. This isn't to say that your research is required to contribute new knowledge to the field; that is generally beyond the scope of a class project. More often, students will end up testing an existing theory, or trying to replicate the results of others. Either way, it's important to understand where our knowledge stands at the time of your research. This will help you to formulate hypotheses and predictions. Your background research should put you on solid footing before starting to design an experiment.

We advise students to rely more on scientific journal articles than on books. The information in books tends to be dated and the amount of information in a book may be overwhelming. If you find a book that relates to your project topic, it is advisable to read selectively, using the table of contents and index to find the relevant sections of the book. For finding scientific journal articles, most students will have access to databases like Web of Science, First Search, ScienceDirect, or SpringerLink. If you're not familiar with your library's system, ask a librarian or your instructor for assistance. Try doing a search using your animal and topic as keywords. Depending on the animal and topic you may need to change keywords to make your search broader or narrower. Reading 10 to 20 scientific journal articles is a good range in finding background information for most class projects.

What is a hypothesis and how do you make one?

As we've said, the first step in a scientific project is to identify an interesting question. The question may result from reading that you have done, from first-hand observations you've made, or may have been suggested to you by an instructor or by this lab manual. For example, you may have watched your dog and observed that it frequently urinates while you're on walks. This leads to the question: Why does my dog urinate when we're on walks?

There are many possible explanations for this behavior. Among the possibilities are that the dog avoids urinating in its own space (your yard), that there is something physiological about walking that stimulates urine production, that urination is a way of scent-marking territories, and so on. To take the possible answers and try to test to determine which answer is mostly likely the correct answer, you need to formulate these explanations as hypotheses.

Remember that a hypothesis is basically a prediction about the outcome of an experiment. A typical animal behavior hypothesis would be of the following general form: Under such-and-such conditions, the organism will exhibit such-and-such behavior. You can then develop a series of more specific hypotheses that relate directly to your experiment: "When exposed to a light source, mealworms will move away from the light."

It's difficult to overemphasize the importance of hypotheses. These provide the framework for your study.

More formally, a hypothesis is a statement of an experimental result that supports an answer to the question. It is usually paired with a null hypothesis, which does not support that answer. When we do an experiment we usually state our expectations about the results in a formal way, as hypotheses. The **null hypothesis (H_0)** is the expectation that there is no effect of whatever we're testing. The **alternative hypothesis (H_1)** is the "positive" result – for example, a difference between a control and an experimental group of animals.

So, taking the dog urination example a step further, you might decide that you can gain insight into the larger question by collecting dog urine and pouring measured amounts at the base of trees. Your hypotheses would then be:

H_0 = The presence of dog urine has no effect on the urination behavior of my dog

H_1 = The presence of dog urine at the base of a tree will change the likelihood of my dog urinating at the base of the tree.

You can see that this hypothesis is much narrower than the original question, and also much more specific than the various answers we listed. The hallmarks of a good hypothesis are that: (1) the hypothesis be directly testable by making observations or designing an experiment and (2) the result of testing the hypothesis will give you insight into the original question. In order to form a picture of what the correct answer to the original question might be, you generally will have to test a large number of hypotheses.

One very important pitfall to avoid is deciding what you think the answer to the question is before you test any hypotheses. Good scientists never set out to "prove" anything. If a scientist tells you that they are working to "prove" a theory, then you should approach their work with caution, as their work likely is biased from the start. One of the intriguing aspects of science is that the results of experiments may be entirely unexpected. Being open to unexpected results, and to modifying established theories if newly collected data does not fit them, is a key characteristic of good scientists.

Some scientists argue that a true hypothesis must be a test of causation, not just of a relationship between variables in uncontrolled conditions. Generally tests of causation require an experiment, in which a single variable is changed while everything else is held constant. This is a reasonable argument, although it is a different concept of hypothesis testing than you would learn in a statistics course. In animal behavior, it is often difficult to maintain precise control over the myriad of internal (physiological and genetic) and external (social and ecological) factors that might affect behavior; consequently tests of causation are sometimes elusive in studies of behavior.

Developing your experimental design

Testing your hypothesis can involve observation, experimental manipulation, or both. The key here is to develop a way to test that it is feasible, without too much expense or time. You should have access to enough individuals for a large enough sample size to properly test your hypothesis. You should measure something that will provide quantitative information to analyze. Martin and Bateson (1993) give excellent advice on experimental design and measurement of animal behavior. Appendix I summarizes common techniques used in animal behavior studies and gives instructions about how these techniques are used; reading Appendix I will help you in designing your study.

A good experimental design isolates one variable and tests that variable while holding all other variables constant. This is difficult, particularly in a field setting, and you will need to be clever in how you design your controls and your manipulations. For many animal behavior studies, the data are obtained by observation and the hypothesis is tested by comparing animals under differing natural conditions. You might wonder if the size of a bird flock affects its efficiency in finding food, and hypothesize that large flocks will find food more quickly. Manipulating flock size would be difficult, but you might easily find flocks of different sizes so that you could build a dataset that makes the comparison you need.

When using field approaches it is still important to attempt to control variables that may confound your efforts to test the effects of a specific variable. For example, you might hypothesize that plants grow more when exposed to sunlight, and test your hypothesis by comparing growth of plants on sunny and cloudy days. It is reasonable to expect that more sunlight, in the long-run, would result in more plant growth. Unfortunately, cloudy days are likely also to be cooler and more humid than sunny days, so any difference in the growth of the plants in your data might be related to temperature or humidity, rather than sunlight. To test sunlight as a **cause** of plant growth, you would need to carefully control all the other variables that might influence growth.

Try to think through every step of your plan in advance. (This is difficult; there's always something you haven't anticipated.) A good method is to make a data collection sheet – this will help you contemplate exactly what you're going to measure and how. You should also make a list of all the equipment you're going to need, including even the most mundane items, like a pen or pencil to write with.

Conducting the experiment or study

Be sure to leave enough time for this part. An experiment always looks straightforward on paper, but in practice there are always snags and unanticipated problems. Animal subjects are not always cooperative, or they behave in unexpected ways. Wild animals may take longer to locate or to make an appearance, or may be less abundant than you had expected. Unexpected contingencies mean that your project will take more time than you planned.

Invariably, as you start work on your project you will get bogged down in the nitty-gritty details of your experiment and may lose sight of just what it was that you were trying to find out. You should frequently refer back to your original hypotheses to insure that your project remains on track, even if you need to modify your experiments as you proceed. Be sure to include sufficient replication in your study so that you may draw statistically meaningful conclusions from your data. Finally, maintain consistency in the methods you use. This may be especially challenging for student groups. Data collected using variable methods or inconsistent units of quantification are extremely difficult to analyze!

Some examples

We hope you'll use your creative powers to come up with a unique project of your own. To help you visualize what might be possible, here are some examples of project topics:

- Do patterns of vigilance in Canada geese change, depending on flock size?
- Does nest attendance differ between Canada geese nests on islands and those on the mainland?
- Is there a dominance hierarchy among birds within a species, or between species, at bird feeders?
- Can I design an experiment to test whether my (dog, cat, fish, other animal) has color vision?

- Does my (dog, cat, other animal) identify me as an individual? If so, what cues does he/she use for identification?
- Can I design an experiment to test whether an animal species is capable of time-place learning?
- Does the color red really serve as an alarm or warning coloration?
- What effect does moving scent marks (a great way to do this is to move "yellow snow" made by dogs) have on territorial behavior of animals?
- What effect do domestic dogs have on the behavior of wildlife? Are wildlife near city or trails more habituated than wildlife in more remote areas?

The final steps of the project, data analysis and presenting your results, are covered in Chapter 2 and Appendix II.

References and suggested reading

Carey, S. S. (2003). "A Beginner's Guide to Scientific Method," 3rd ed. Thomson, Belmont, CA.

Gauch, H. G. Jr. (2003). "Scientific Method in Practice." Cambridge University Press, New York.

Martin, P. and Bateson, P. (1993). "Measuring Behaviour," 2nd ed. Cambridge University Press, New York.

Chapter 2
Presenting your results

This chapter starts where Chapter 1 left off. Science does not end when an experiment is completed; scientists must put their work on public view, available for assessment and evaluation by other scientists. Publication of results allows the information you have gathered to be incorporated into the general body of scientific knowledge. Publication also allows other scientists to critique your work and to attempt to replicate your experiments.

For students in laboratory classes, sometimes reporting data in laboratory reports becomes a routine task that simply completes a "cookbook" project. We have attempted, in the labs in this manual, to give you the opportunity to design your own hypotheses and to make your own experimental choices. Along with this comes the opportunity to report your results in either oral or written form. This chapter serves as a beginning guide for these reports.

The context for presenting your results will depend on the structure of the course. If time permits, it is nice to have a "research symposium" where students can share their results with each other, either as posters or oral presentations. Whatever be the context, keep in mind that sharing one's results is one of the most important steps of the scientific process. If you don't share what you have learned, then the larger community gains nothing from all your efforts.

1 Describing your data

Once you've completed data collection you may feel like the hard work is done. But one of the most important steps still lies ahead. You have to use your data to test your hypotheses. Do your data support your hypotheses? Occasionally, you can answer your question with a "yes" or "no" simply based on whether the behavior you were looking for occurred or not. More often, you will have to run some statistical tests to see whether your results are meaningful. We've provided some guidance for doing statistical tests in Excel (Appendix II). There are many other software packages available to do statistical tests; consult with your instructor to help you carry out the appropriate analyses.

II Writing a short research paper (lab report)

Clear, written communication of scientific results is essential. Your instructor will specify the length of the writeup. Use double spacing when printing and a font that is large enough for easy reading but not so large it looks like you're trying to get by with too little work. Generally, a typeface like Times or Geneva in a 12 point font is good. Try to write as if you enjoyed doing the experiment and you are excited about the results.

Follow the general format of Introduction, Methods, Results, Discussion, and Literature Cited. In the Introduction, provide some basic background information for the question you are testing. What is/are your hypothesis(es)? Avoid simply regurgitating the text of the lab manual. Next, in the Methods section, describe the experimental design well enough that an intelligent person could replicate your experiment, based on your description. A diagram here could be very helpful. In the Results, describe the findings of the experiment you performed. Include the results of statistical analyses of the data and a graphical representation of the data. Your instructor may provide you with a summary of all the students' data, so that you have a larger sample size for your analyses. Interpret your results in the Discussion section. How do your results support or not support your hypotheses? What do your results mean in the context of other similar studies? You should cite at least three papers from the primary literature in your discussion (ask your instructor if you are not sure how) that provide a frame of reference for your work. Are there other published studies that found a different result? Has a similar study been done with a different organism? After the discussion, include a Literature Cited section. Use the Literature Cited section of a published paper in the scientific journal *Animal Behaviour* as a formatting guide.

Worksheet 2.1 is an outline of points that a reader will probably assess when grading your lab report. You can use the Worksheet as a checklist when you're writing the report.

There are two purposes for writing a lab report assignment. First, it gives you valuable experience in presenting data you collect in a short written form. Second, you will be required to dip into the primary literature and learn how to place your results into an appropriate scientific context. Your instructor should be more than willing to help with statistical analysis, primary literature searches, and with reading drafts.

III Giving an oral presentation of lab or project results

Oral presentations follow the same outline as written presentations. Structure your presentation with an Introduction, a Methods section, then Results, and a Discussion. You should give relevant citations to the scientific literature, as well. Worksheet 2.2 is a sample grade sheet for an oral presentation. You may be asked to turn in an Abstract, a short written summary of your project, along with your oral presentation; the Abstract is included in the Worksheet. As with the written lab reports, the Worksheet is an excellent checklist as you prepare your presentation.

The advantage of an oral presentation is that you will find it easier to bring your enthusiasm for your work into your presentation. If you are nervous about making presentations in front of your peers and your instructor, the keys are preparation and practice. If you haven't given many presentations, make detailed notes, or even write out a complete script of what you want to say. Then practice (inflict the presentation on friends or family) until you are confident enough not to read the entire talk from your notes. Remember that many performers on television, who appear to be organized, spontaneous, calm, and collected, are actually reading scripts from teleprompters – you needn't feel inadequate if you need to work from a script.

Most talks these days are done using Powerpoint visuals, and likely your talk(s) in this class will involve Powerpoint slides. Here are some Powerpoint tips:

1. Use your slides to emphasize important points, not to convey your entire script. Avoid using the templates that come with the Powerpoint program, as you'll end up with lots of bulleted lists, and a boring talk.

2. You aren't stuck with only using the computer presentation. If appropriate, bring in your experimental animal, your experimental apparatus, or other relevant materials so your audience can see what you actually worked with. If a picture is worth a thousand words, the actual animal or apparatus is worth ten thousand.

3. Keep the slides visually simple, so that the viewer's eye is drawn to the key point you're trying to make. Use animations sparingly, as they are generally more distracting than helpful.

4. You'll be tempted to grab images from the web, such as pictures of your study organism. Make sure you credit image sources and respect copyrights, as you would for any intellectual property.

5. Remember, above all, that the visuals are to support what you have to say, rather than the other way around.

6. As a general rule of thumb, you shouldn't use more than one slide per minute. If you are allotted 10 minutes, then about 10 slides is appropriate. If you have organism pictures that you can go through quickly, you might have a few more slides than this, but if you have a slide of a method that requires complicated explanation, you'll need to use fewer slides.

IV A final word on reporting your results

Above all, remember that you learned something interesting doing the experiments or observations, and reporting those results is critical to the advancement of science. Experimental results that go unreported ultimately do not contribute to the scientific enterprise; while the researcher may have enjoyed the experience of the study, the work only has a lasting impact if the results are communicated.

A semester research project in a college class may or may not yield statistically significant results, but the process of communicating these results in written and oral form is absolutely invaluable for any student wishing to pursue a career in any branch of science. The substance of your presentation is paramount; however, a clear and enthusiastic presentation style will help maintain the attention of your audience. This is more difficult in writing than in oral presentations, but drawing your audience to you is the key to successful presentations, regardless of the format.

Worksheet 2.1 Sample animal behavior lab report grading sheet.

Title

_____ The title specifically indicates the main point of the study

Introduction

_____ Clear statement of the specific questions addressed

_____ Context is provided (background information, likely from scientific literature)

_____ Hypothesis stated

_____ Null hypothesis stated

_____ Predictions stated

_____ Hypotheses are stated as _candidate explanations_ for why the predicted trend is expected to occur

_____ Rationale provided to explain why the question was addressed and to justify your hypothesis

Materials and Methods _(write in past tense)_

_____ Design of study or experiment is clear and complete (it can be brief, but write in paragraph form)

_____ Data analysis techniques are included as a method (calculations made, statistical tests used)

_____ Data analysis technique is appropriate for the dataset

Results _(write in past tense)_

_____ Main trends are summarized, but implications are not discussed in this section

_____ All general statements are supported with reference to data and/or the results of your statistical analyses (i.e. p values, R^2 values, etc. If referring to a trend shown in a graph, give the figure number of the graph)

_____ Graph type is appropriate for the dataset

_____ Graphs formatted correctly, using Excel (include meaningful title/caption, capitalization, axis labels, units)

Discussion

_____ Main conclusions are clearly stated and are related directly back to the original hypothesis

_____ Data trends and statistical significance are correctly interpreted

_____ A scientific interpretation of the results is provided: what is a _biological_ explanation for your results?

_____ The results are discussed in relation to your knowledge of the scientific literature on the topic. (Do your results support/contradict the findings of other studies? If they contradict, explain why you think your results are different from that expected. Is your study an exception? Did you use different methods?)

_____ The possible theoretical or practical applications of the results are presented

_____ Discussion ends with a cohesive, concluding paragraph that places the specific results of this study into a larger, real-world context

Grammar and Readability

_____ Grammar, punctuation, and spelling are correct

_____ Tone is professional and authoritative

_____ Main point is logically developed from sentence-to-sentence and from paragraph-to-paragraph

Literature Cited

_____ Citations are provided in the paper whenever referring to another person's work or ideas

_____ At least three primary references are incorporated into the body of the paper

_____ Citations formatted correctly in text

_____ Citations formatted correctly in Literature Cited section

_____ Literature Cited section contains all references that were cited in text, none that weren't

Worksheet 2.2 Sample semester project grade sheet.

Presentation

Title Slide

_____ Student's name

_____ Title

Introduction (usually 3–5 slides)

_____ Explanation of main concept/theory tested

_____ Hypothesis

_____ Necessary background information of study system (refers to primary sources)

Methods (usually 2–4 slides)

_____ Explanation of experimental setup

_____ Explanation of data collection methods (including statistical analyses)

Results (usually 2–4 slides)

_____ Graph or chart type is appropriate for the data set

_____ Graphs or charts are accurate and complete

_____ Indication of trends/significant results (including reference to p value)

Discussion (usually 2–3 slides)

_____ Correct and thorough interpretation of results

_____ Results support/do not support original hypothesis(es)

_____ Discussion of how the results of this study fit into the broader context of the topic (must refer to a total of at least three primary sources in introduction and discussion)

Overall Presentation Quality

_____ Organization/clarity

_____ Quality of visual aids

Written Abstract (if required by instructor)

_____ Introduction to main concept

_____ Hypothesis(es)

_____ Statement of findings

_____ Brief interpretation

_____ References cited correctly in body and at the end (includes at least three primary references)

Section 2
Field Studies in Animal Behavior

Chapter 3
Dominance-discovery trade-offs

Behavioral ecology is the study of how abiotic and biotic environments affect animal behavior (Krebs and Davies, 1997). One of the goals of behavioral ecology is to understand how natural selection operates on the costs and benefits of animal behaviors, and how the interactions among costs, benefits, and the environment ultimately result in the behaviors we observe. The trade-offs between the costs and benefits of particular behaviors result in variation in behavioral strategies, both among individuals of a species and among different species.

Many behaviors can be understood as trade-offs between competing interests or abilities. Brown and Kotler (2004) give an interesting discussion of the trade-offs between the risks that foragers encounter when looking for food and the benefits of finding food. Not surprisingly, for animals, the risks of being injured by potential prey or of becoming a prey item play large roles in shaping their foraging behavior. For example, Pomeroy et al. (2006) manipulated the amount of cover for foraging sandpipers. They found that sandpipers adjusted their foraging behavior along shorelines depending on their perception of the danger of being preyed upon; they spent less time in the less-sheltered control transects than in the sheltered treatment transects.

Another foraging context in which trade-offs can be observed is competition for dietary resources (Savolainen and Vepsalainen, 1988). This is a particularly interesting and common interaction among animal species. Differing competitive strategies among species help explain how several species can coexist in a habitat even when their dietary needs significantly overlap. In this lab, we'll take a look at an ecological system in which behavioral trade-offs in competitive style during foraging are apparent.

Ground-foraging arthropods, such as ants, beetles, and spiders, are a good example of a diverse animal assemblage in which species show a large degree of dietary overlap (Fellers, 1987; Davidson, 1998; Holway, 1999). These small animals actively search their environment for food resources. Among ground-foraging arthropods competition for food is intense, and many species are easily attracted to baits (Davidson, 1998; Holway, 1999). We'll use this attraction to baits as a starting point for contrasting competitive styles among species.

How might strategies differ among ground-foraging arthropods? Speed of discovery of the food could be a critical factor. When time is of the essence, ecologists refer to the process as "scramble competition" (Brannstrom and Sumpter, 2005; Snaith and Chapman, 2005). This strategy means excelling at finding food as soon as possible after it becomes available; these animals are good "discoverers". Another strategy might be to seize control of a food resource from a competitor, and to guard the resource against subsequent competitors; animals with this ability are considered "dominant". Good discoverers may sacrifice strength and fighting ability to be fast moving while dominant animals may be slower moving but better fighters; this is a key trade-off in the behavior of ground-foraging arthropods.

How does the trade-off between these two strategies play out in real life? Is it possible that, contrary to expectation, an animal that discovers a resource quickly also defends it against competitors? We will investigate these questions by quantifying the behavior of the ground-foraging arthropods that find our food baits.

I Goals

- To learn how to use transects and baits in the field as a method for studying foraging behavior
- To observe food procurement strategies which differ among potential competitors
- To relate these strategies to trade-offs in foraging behavior and to the evolution of feeding and defensive adaptations

II Questions and hypotheses

In this lab, we investigate the question of whether there is a trade-off between bait discovery and bait dominance. Informally expressed, the hypothesis is that some species will quickly discover the bait, but then they will be displaced by more dominant species that are slower to find the bait. We also hypothesize that behavioral interactions between species, such as fights or avoidance, will facilitate dominance of the bait by certain species.

▪ III Methods

Your instructor will have selected a large area, perhaps a park or woodland, where you will place your baits. Ideally, there will be some topographic variation (hills and valleys) and shaded areas. Establish a 100 m long transect – a straight line used for orientation – and identify bait stations every 10 m along the transect. At each bait station, place bait cards on the ground so that the entire perimeter of the card is in contact with the ground (this will allow small arthropods to walk onto your card rather than under it). Bait cards consist of a 3 inch × 2.5 inch (7.5 cm × 6.25 cm) white notecard; these can be created by cutting a 3 inch × 5 inch (7.5 cm × 12.5 cm) notecard in half. The white card provides a high-contrast background and will help you observe small animals that would otherwise be difficult to observe. Your instructor may provide you with the bait to be used, or you may wish to experiment with a few different types. A few good baits include canned tuna in water (drained), Pecan Sandies™, or whole frozen crickets (thawed). Use only a small amount of bait about 1 cm in diameter. Following the example in Worksheet 3.1, use Worksheet 3.2 to make a diagram of your research area and to record how the transect was laid through the area, the positions of the baits, and insects at each bait location. At active bait stations, you may need two people per bait card, an observer and a notetaker.

Before starting, take a few moments to assess and make notes on the conditions of the surrounding environment. Be sure to include conditions that may be important to the focal animals! For example, soil moisture may be more important to these animals than relative humidity of the air. If this were a larger project, you would develop a sense for the effects of these conditions. Place the baits on the cards and start the timer. Watch the baits, recording the time and identity of foraging arthropods. Record an accurate description and count of each visiting species. Try, with the help of your instructor, to identify the organisms to family, or perhaps to genus. You can also keep sample specimens (vouchers) for subsequent identification, with the help of your instructor, to subfamily or genus.

It is very likely that at some point, visitors will become too numerous and frequent to keep continuous track of them all. Rather than keeping track of the arrival and departure of each individual arthropod, a good option to deal with this is to collect an instantaneous sample at predetermined time intervals, such as every 5 minutes after the initial discovery of the bait. An instantaneous sample is one in which the activity of all the animals present is recorded at a single time. If there are a large number of animals, either multiple observers or a photograph that can later be analyzed are good ways of obtaining instantaneous counts.

While counts of animals may show differences between species that are good discoverers and others that are dominators, behavioral observations of encounters between species will help you to understand the mechanisms by which one species might displace another from the bait. Watch, as closely as you can, to see if there are direct physical contacts or fights between species. Do some species seem to avoid contact with others? Record your observations on Worksheet 3.2.

While making these observations, try to avoid disturbing the animals. Many of these small animals have acute vision, so sudden movements may scare them; others may be startled by the CO_2 in your breath, so avoid breathing on the bait; all will be terrified by physical disturbance, so do not bump the bait or the bait card!

IV Interpretation

Your small group should assemble your results so that the data from the entire class can be summarized and prepare for the class discussion by answering the questions in Worksheet 3.3.

References and suggested reading

Brannstrom, A. and Sumpter, D. J. T. (2005). The role of competition and clustering in population dynamics. *Proc. Roy. Soc. B-Biol. Sci.* **272**, 2065–2072.

Brown, J. S. and Kotler, B. P. (2004). Hazardous duty pay and the foraging cost of predation. *Ecol. Lett.* **7**, 999–1014.

Davidson, D. W. (1998). Resource discovery versus resource domination in ants: a functional mechanism for breaking the trade-off. *Ecol. Entomol.* **23**, 484–490.

Fellers, J. H. (1987). Interference and exploitation in a guild of woodland ants. *Ecology* **68**, 1466–1478.

Holway, D. A. (1999). Competitive mechanisms underlying the displacement of native ants by the invasive Argentine ant. *Ecology* **80**, 238–251.

Krebs, J. R. and Davies, N. B. (Eds.) (1997). "Behavioural Ecology: An Evolutionary Approach," 4th Ed. Blackwell Science, Oxford, Malden, MA.

Pomeroy, A. C., Butler, R. W., and Ydenberg, R. C. (2006). Experimental evidence that migrants adjust usage at a stopover site to trade off food and danger. *Behav. Ecol.* **17**, 1041–1045.

Savolainen R. and Vepsalainen, K. (1988). A competition hierarchy among boreal ants: Impact on resource partitioning and community structure. *Oikos* **51**, 135–155.

Snaith, T. V. and Chapman, C. A. (2005). Towards an ecological solution to the folivore paradox: patch depletion as an indicator of within-group scramble competition in red colobus monkeys (*Piliocolobus tephrosceles*). *Behav. Ecol. Sociobiol.* **59**, 185–190.

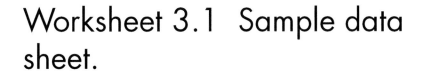

Worksheet 3.1 Sample data sheet.

Start Time: 2:00 pm   Bait Type: <u>Pecan Sandie and Sugar Water</u>
Location and description: <u>Open field. Moist, sandy soil with grasses and forbs.</u> Air: <u>33°C, soil surface: 28°C, mostly sunny, light breeze.</u>

Time	Arthropod Identity	#	Notes
2:02	Metallic green fly	1	
2:07	Metallic green fly	4	Mating observed
	Dark gray, fuzzy fly	1	
	Roach	1	<u>Ewww!</u>
2:12	Metallic green fly	6	
	Dark gray, fuzzy fly	2	
	Small red ant	1	Touched bait with antennae and ran away!
2:17	etc…		
etc…			

Worksheet 3.2 Dominance and discovery data sheet.

Start Time: Bait Type:
Location and description:

Time	Arthropod Identity	#	Notes

Worksheet 3.3 Dominance and discovery discussion questions.

Your group should discuss answers to these questions, which then will be used as the basis for a discussion by the class.

1. Can you think of differing strategies animals use to procure food, and the trade-off between costs and benefits associated with these strategies? (e.g. sit and wait versus active hunting; filter feeding versus selective feeding; etc.)

2. What was the identity of the first animal to discover your bait?

3. Describe any morphological or behavioral adaptations that might allow this organism to be an adept resource discoverer. Did this animal maintain possession of the resource for the whole time? If so, describe any behaviors it exhibited to protect its resource from competitors.

4. How long, on average, did it take animals to discover your group's baits? If you had time, you may have sampled more than one habitat. Was there a difference in discovery time among habitats? If so, why do you think such differences might be present?

5. If another animal took over the bait, what behaviors did it exhibit? What was the identity of the animal that eventually dominated your bait? Describe any morphological or behavioral adaptations that might allow this organism to be an able resource defender.

6. Finally, was there a trade-off between dominance and discovery at your bait? In other words, are some kinds of animals better at discovering baits and others better at dominating baits?

Chapter 4
Pollination behavior

Plants and their pollinators are mutually dependent. Plants require the services of pollinators in order to reproduce, and pollinators depend on the food rewards provided by the plants for their survival (Waser and Ollerton, 2006). Plants evolve to maximize their chances of fertilization, while their pollinators evolve to maximize the amount of food collected. The strategies used by plants affect the evolution of their pollinators and vice versa. In this introductory section we discuss the complexities of the evolutionary interactions between plants and their pollinators. You are then given the opportunity to study pollination biology and to compare different observational techniques used by animal behaviorists (Kearns and Inouye, 1993).

▪ 1 The plant's "point of view"

Plants have evolved a variety of ways of attracting pollinators and ensuring that pollen is carried from one plant to another. These include fragrances, showy displays of color, and rich food rewards. Floral specificity, or specialization of a pollinator species on a flower species, enhances pollen exchange for the plant species. Much of pollinator/plant coevolution seems to revolve around plants attempting to enforce this kind of fidelity (oligolecty) (Blarer et al., 2002). Plant species may evolve elaborate flower structures that only certain insects or birds can exploit. For example, a long, narrow corolla (petals that are fused together) may allow only hummingbirds, with their long bills, to collect nectar from a plant species. Plants whose flowers are accessible to only one or a few animal species limit competition among animals for nectar and pollen, thereby rewarding the pollinators for their fidelity (Schulke and Waser, 2001).

II The pollinator's "point of view"

Pollinators move among plants in ways that minimize the amount of time and effort spent to obtain food. They are able to learn to extract food efficiently from different types of flowers, and to avoid flowers that are not rewarding. Pollinators have no direct interest in ensuring pollination of a plant species, and if more than one species of flower is easily accessible, pollinators will switch among plant species as they move, defeating the plant's purpose of obtaining pollination services (Thompson, 2001). Some bees even learn how to evade the devices plants have that attempt to enforce fidelity by chewing holes in the corolla of deep flowers to gain entry to the nectaries without contacting the anthers or stigma. Behaviorally, the challenges for animals visiting flowers are to learn to obtain the rewards from each flower species, to avoid unrewarding flowers, and to collect the most food with the least effort and risk of being preyed upon.

As you watch pollinators in this lab, keep in mind the disparity in evolutionary "points of view" between the plants and animals involved in these relationships. Can you identify features of animal behavior that are detrimental to the pollination of the plants? Can you see features of the plant that might force pollinators to restrict their visits to that plant species?

III The kinds of pollinators

Most insect pollinators are members of four orders, the beetles (Coleoptera), the flies (Diptera), the moths and butterflies (Lepidoptera), and the ants, wasps, and bees (Hymenoptera). In addition to insects, other important pollinators are birds, particularly hummingbirds, and some bats.

Diptera: The true flies, or Diptera, are characterized by having a single pair of wings on the middle segment of the thorax (Figure 4.1). Instead of hind wings, they have a pair of knob-like structures, called the halteres, which function like gyroscopes in keeping the fly oriented in flight. Flies are common visitors to flowers and may be important pollinators. For example, the Marsh marigold, *Psychrophila leptosepala*, a common subalpine plant found in marshy areas, is probably primarily pollinated

Figure 4.1   A fly on the subalpine plant, marsh marigold (photo by Michael Breed).

by flies. Some flowers, notably, arums, members of the jack-in-the-pulpit family are adapted for pollination by flies that normally visit carrion and dung; they produce malodorous (to us) scents to attract their pollinators. These species occur primarily in the tropics.

In addition to the characteristic of having one pair of wings, which can often only be seen on close observation, flies seen on flowers usually have short antennae that may be club-like in appearance. Flies are much more likely than bees to remain for extended periods on a flower and are also more likely to hover in the air around flowers. Some flies appear near flowers because they lay eggs on the flowers; the larvae that hatch from these eggs are parasites of other insects that visit the flowers.

Hymenoptera: Hymenoptera include the wasps, bees, and ants. Most of the Hymenoptera that you will see have a distinct constriction between

Figure 4.2 A bumblebee, resting between stops on flowers. Bees sometimes pause to move pollen from their thorax to their hind legs or underside of their abdomen (photo by Michael Breed).

the thorax and abdomen, giving them a "waisted" appearance. Many female Hymenoptera have a sting, which is used to paralyze prey or in self-defense; males do not possess stings. Hymenoptera have two pairs of clear wings and their antennae are usually about the same length as the front legs. The antennae have the appearance of beads on a thread, although the beading is not pronounced.

Bees are the champion pollinators because they rely on flowers for carbohydrate food (nectar) and protein food (pollen) (Figure 4.2) (Cane, 2001). This reliance causes them to visit flowers on a regular basis. Bees have special adaptations for pollen collection that enhance their pollination ability. First, the hairs on their body are branched, so that the pollen grains stick easily to their hairs. This branching can be seen under a dissecting microscope. Second, they have mats of hairs on their hind legs

Figure 4.3   A beetle (photo by Michael Breed).

or on the bottoms of their abdomens for carrying pollen. Bees see ultra-violet, blue, and yellow as their primary colors, but don't see red as a color. For this reason, "bee flowers" are usually blue, yellow, or white and often have ultraviolet color patterns in them that we can't see.

Most bees are solitary, with a single female living in a nest, in a twig, or in the soil. Social bees include the bumblebees, some sweat bees, and the honeybee. The honeybee is not native to the Americas, but now occurs in nearly all North American ecosystems.

Coleoptera: The beetles are easily identified by their hardened forewings, which cover the clear hind wings and their abdomen (Figure 4.3). These forewings, or elytra, protect beetles, giving them a shell-like covering. In flight the forewings are held out, but do not propel the insect; the hindwings have this job. There are more species of beetles than any other kind of insect, with over a quarter of a million known to science. Beetles occupy almost every conceivable ecological niche except the oceans. Beetles visit flowers for a variety of reasons. Some species lay eggs on the flowers; then, as larvae, the beetles consume the plant's seeds. Other species eat pollen and nectar. We often see blister beetles on flowers late in the season; they use the outposts for mating.

Lepidoptera: Butterflies and moths feed as larvae on plant tissue and, as adults, rely on nectar as an energy source. They are easily recognized

Figure 4.4 A butterfly, collecting nectar from a flower. Note its proboscis, or feeding tube, which extends from the front of the head into the flower; this is used to suck nectar from the flower (photo by Michael Breed).

by their scaly wings. Butterflies typically have thread-like antennae, while moth antennae are usually adorned with long hairs that allow the moths to sense the pheromones that their mates produce.

In many parts of the United States, it is common to see yucca plants, which have a special relationship with a moth (Powell, 1992). Females of this moth species collect pollen from one yucca plant and carry it to another plant, where they pack it onto the pistil. The moth's legs and mouthparts are specially adapted for this task. This behavior ensures pollination of the yucca and production of seeds. The moth then lays eggs in the developing seedpod, and the moth's larvae consume many, but not all, of the seeds. The yucca's relationship with the moth is a classic example of coevolution.

Butterflies are common daytime visitors to flowers (Figure 4.4). Moths more often fly at night and therefore attract less attention. Hawkmoths are particularly important pollinators at high elevation because of their ability to shiver to raise their body temperature; this allows them to fly during cool mountain mornings and evenings.

IV Plant structures

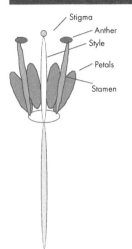

An insect- or bird-pollinated flower usually has brightly colored petals surrounding the reproductive parts of the flower. This simplified diagram (Figure 4.5) shows the basics of flower structure. The stigma is the receptacle for the pollen, which is produced in the anthers. Nectaries can occur in a variety of locations, but are usually found at the base of the petals. More complicated flower structures evolve when petals are fused together or are different sizes and shapes on the same flower.

Figure 4.5 A simplified diagram of a flower.

V Goals

- To learn to identify the major insect pollinator orders
- To learn some basic scientific techniques in the study of pollination biology
- To practice using focal animal and scan sampling techniques
- To understand the evolutionary dynamics which have shaped the plants that pollinators visit and pollinator behavior

VI Questions and hypotheses

Your instructor will probably give you some latitude in designing your hypotheses. Your instructor will have chosen, prior to the lab, a field area in which plants are flowering and pollinators are flying. This could be ornamental plantings on your campus, weeds in a vacant lot, or a more pristine meadow; any of these settings will allow you to make interesting discoveries. We suggest that you spend a little time – 15 to 30 minutes should be adequate – observing pollinators in action before specifying your hypotheses. Making these preliminary observations will help you to design hypotheses that can be reasonably tested. Here are some interesting sample questions to think about as you make your observations:

- You've chosen a plant species to watch. Does the flower seem to be relatively simple or relatively complex? Can you test a hypothesis that relates floral complexity to pollinator specialization?
- Pollinators may need to learn how to handle each plant species they visit. If this learning takes time and energy, you might hypothesize that it would be best to pass over unfamiliar flower types, even if the pollinator has to fly some distance to find a flower type it has learned to exploit.
- It is probably a waste of time and energy for a pollinator to visit a flower that has recently been visited by another pollinator. Can you test the hypothesis that recently visited flowers are avoided?

Of course, you should be able to come up with other really good questions of your own, based on your observations of pollinator behavior.

VII Observational strategies

We'll collect data on pollinators using three techniques that are commonly used by pollination biologists. We describe the actual procedures in Worksheets 4.1–4.3 which accompany this lab. Two of these techniques involve observation of focal organisms. In the first focal sample, you'll choose a flower and watch for pollinator visits. In the second focal sample, you'll choose an animal that is visiting flowers and follow its visits to different flowers. Comparing your results from these two techniques will give you a feel of how plant and animal "points of view" differ in pollination relationships. The third technique uses a scan sample to survey pollinators on a plant species (see Appendix I for more detail on the differences among these sampling techniques).

VIII Interpretation

Some students in your class would have used focal sampling, some scan sampling, and some would have used both. These are key techniques in animal behavior. Discuss, as a class, the relative strengths and weaknesses of focal and scan sampling. When might one be a better approach than the other?

Share the results of your observations to your class as a Powerpoint presentation (see Chapter 2 for more details on how to prepare your presentation). Be sure to include your observations on the adaptations of plants and insects involved in pollination relationships, and your conclusions concerning the hypotheses you set out to test.

References and suggested reading

Blarer, A., Keasar, T., and Shmida, A. (2002). Possible mechanisms for the formation of flower size preferences by foraging bumblebees. *Ethology* **108**, 341–351.

Cane, J. H. (2001). Habitat fragmentation and native bees: a premature verdict? *Conserv. Ecol.* [Online] URL: http://www.consecol.org/vol5/iss1/art3.

Kearns, C. A. and Inouye, D. W. (1993). "Techniques for Pollination Biologists." University Press of Colorado, Boulder, Colorado.

Powell, J. A. (1992). Interrelationships of yuccas and yucca moths. *Trend Ecol. Evol.* **7**, 10–15.

Schulke, B. and Waser, N. M. (2001). Long-distance pollinator flights and pollen dispersal between populations of *Delphinium nuttallianum. Oecologia* **127**, 239–245.

Thompson, J. D. (2001). How do visitation patterns vary among pollinators in relation to floral display and floral design in a generalist pollination system? *Oecologia* **126**, 386–394.

Waser, N. M. and Ollerton, J. (Eds.) (2006). "Plant-Pollinator Interactions: From Specialization to Generalization." University Of Chicago Press, Chicago.

Worksheet 4.1 Watching a flower (focal observations).

Instructions

Choose a flower species to watch and find a comfortable vantage point. Note the time and start a 30-minute observation period. Each time a pollinator visits the plant or plants that you can see, make a notation and to the best of your ability identify the pollinator.

Plant species (if known) —————————————————————

Color of flower(s) ——————————————————————————

Typical number of flowers per plant —————————————————

Typical number of plants in clump ——————————————————

Conspicuousness – from how far away can you see the plant or clump of plants? ——————————————————————————

Time (from start) of visit	Identity of Pollinator

Interpretation

Based on your data, do you think your flower species attracts specialist or a generalist pollinator? Why?

Worksheet 4.2 Following a pollinator (focal observations).

Instructions

Find a likely pollinator – a butterfly, bee, beetle, or fly, for example, and follow it, placing a flag each time it stops on a flower. Construct a crude map on the grid below, showing the animal's movements. Use letters to designate different plant species at which the pollinator stops. Connect the stops with a line. If the pollinator specializes in one or two species, write letters on the map to indicate locations of plants that were not visited, but do not connect them to the line showing the pollinator's movements.

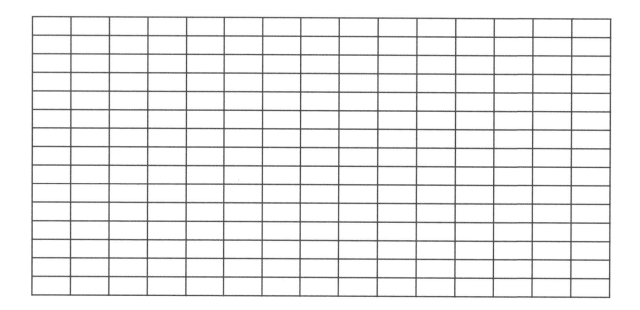

Interpretation

Based on your data, do you think your pollinator specializes on a particular plant species?

Worksheet 4.3 Scan samples.

Instructions

Choose a plant species. Walk a straight line through the meadow. For each plant or clump of plants record how many pollinators of what type you see. Stop when you reach the edge of the meadow or run out of plants.

Plant species (if known —————————————————————————

Color of flower(s) ————————————————————————————

Typical number of flowers per plant ——————————————————

Typical number of plants in clump ———————————————————

Conspicuousness – from how far away can you see the plant or clump of plants? ————————————————————————————

Clump #	List of pollinators seen

Interpretation

Based on your data, do you think your flower species attracts specialist or a generalist pollinator? Why?

Chapter 5
Foraging

In the previous two labs we've investigated the behavioral ecology of foraging. In this lab you'll be given the opportunity to act as a forager yourself, by playing one or more games. By being actively involved in the behavior you gain a much better intuitive understanding of the decisions that animals make while foraging. For animals, strategies for acquiring food may vary as widely as the organisms themselves; these games will give some experience in the diversity of strategies available to foraging animals.

Foraging is a large topic in animal behavior. Kamil et al. (1987) provided a scientific overview of foraging behavior that helped to shape research on foraging over the last 20 years. Krebs and Kacelnik (1991) focus on decision-making in behavior, including foraging. An overview of behavioral ecology is given by Krebs and Davies (1997).

In this lab there are three options, all of which call on the students in the class to act as foragers. Your instructor may divide you into groups to explore each option, or choose one or more of the options for the entire class to pursue.

I Herbivore–predator interactions in a system with three trophic levels

For today's exercise, it is useful to divide animals into two main groups on the basis of their eating habits. Herbivores are animals that primarily eat plants; predators are primarily meat-eaters (i.e. they may eat herbivores). Other eating habits are important of course, but omnivores and decomposers, for example, are not relevant to the current exercise. On a basic level, it is obvious that a herbivore will have a very different relationship with its food (which does not move around) than a predator whose food must be caught and killed. Often, herbivores must move around their habitat, seeking patches of their preferred plant resources. Herbivores typically spend more time finding and consuming food than on any other activity. Predators, on the other hand, must find a mobile food source, then catch and kill it. The typical predator does not eat constantly, but gorges on a kill and then may not hunt again for some time. Therefore, a much smaller percentage of a predator's time is spent looking for food, but finding and obtaining the food is more energy intensive and riskier.

Foragers, be they herbivores or predators, have developed a variety of adaptations that improve their success at obtaining food. These adaptations often involve the enhancement of certain sensory organs that allow them to perceive their food (smell, sight, etc.). An example of a behavioral adaptation of this sort is the social organization of wolves. While even a single wolf is a formidable hunter, a wolf pack can kill much larger prey than a single wolf alone and the kill is shared among all. What are some other adaptations in foraging behavior that might improve an animal's success in finding food?

While all animals must eat to survive, their "food" usually does not want to be eaten. Being eaten generally inhibits reproductive output and thereby reduces fitness of the prey item. Therefore, there is a strong selection pressure on traits that reduce the likelihood of becoming food to another organism's meal. This is true for both plants and animals. For example, plants may produce chemical defenses against herbivory, and many animals have developed some sort of camouflage to prevent being seen by predators. A great many of the behaviors we see in animals, such as herding, flocking, and vigilance, are related in some way to minimizing individuals' vulnerability to predation. What are some other behavioral adaptations of this sort?

If foragers are developing better adaptations for finding or capturing their food, and their food is developing better adaptations for evading detection or capture, what happens? Effectively, foragers and their food are locked in an arms race in which selection favors better defenses against being eaten in the prey, as well as better ability to circumvent these defenses in the consumer. In this exercise, students, playing the roles of herbivore and predator, will have the opportunity to participate in this arms race first-hand, each adapting their strategies to better survive the selection pressures imposed by their environment.

A Goals

- To gain a first-hand experience of behavioral adaptation and the types of selection pressures experienced by animals in the wild
- An understanding of the strategies of foraging, predation, and predator-avoidance that are observed in animal behavior and why they work

B Questions and hypotheses

Given the outline of the exercise described below, students should think of some strategies that could be adopted by the predators and/or the herbivores that might increase their success in the game. They will have opportunities to try out these strategies in the game. For example, herbivores may decide to "herd" together for defense or to appoint individuals to watch for predators. Predators may also develop some sort of cooperative strategy.

C Methods: Macaroni foragers

Your instructor will have selected an outdoors area, preferably a relatively flat playing field, where you will play the predator–prey game. Macaroni noodles will have been scattered over the field, except for one corner. Students are designated "herbivore" or "predator". Usually there should be no more than two predators for a group of 15–18 herbivores. Herbivores forage by collecting the macaroni, while predators hunt and "eat" the herbivores by capturing their flag. When a predator catches a herbivore, it is out of the game for that round and it must surrender all of the macaroni it has gathered to the predator. After catching a herbivore, a predator must wait one minute before hunting again (handling and digestion time). The corner of the field that has no macaroni is a "safe zone" for the herbivores. This is analogous to a den or nest in the wild. Since there is no macaroni within this zone, herbivores cannot survive merely by staying there to avoid being eaten.

The individual with the most macaroni at the end of a round is the winner of that round. If several were working as a group, the total macaroni holdings of the group must be divided by the number of individuals in the group.

The first round is played with all participants acting as individuals. A round should be 5–10 minutes, but can be ended at the discretion of the instructor at any time (e.g. if all of the herbivores have been eaten).

For each successive round, herbivores and predators may hold a strategic planning session beforehand. Strategies decided upon during this meeting should be maintained throughout the following round. Individuals may also decide to employ their own strategies without engaging in a group strategy.

Macaroni is re-scattered between rounds. Roles may also be re-assigned between rounds.

Materials

Macaroni noodles, flags, bags in which herbivores may put their harvest. All participants should wear comfortable clothes and running shoes.

D Interpretation

Record your data on Worksheet 5.1 and write answers to the questions. The most important thing to think about after doing this exercise is how your behavior as a forager reflects the difficulties that foraging animals have in nature. Food is hard to find, and foraging carries with it the risk of becoming another animal's prey. Experience shapes foraging in two ways. First, over long timescales evolution selects against animals that cannot find food or that easily become prey. Second, within an animal's lifespan, learning can dramatically improve a forager's performance.

II The disc equation and decision rules in foraging

Foraging animals often encounter their prey at varying abundances in the environment; some resource patches may be very rich, while others may be depauperate. Therefore, one might expect the rate at which a consumer feeds on its prey to be related to the abundance of the food items. The relationship between prey abundance and the number of prey consumed by a predator is referred to as the *functional response*.

Take a moment to consider how prey consumption might be related to prey abundance. As prey abundance increases, what do you expect to happen to the frequency with which the consumer eats the prey? In Figure 5.1, draw the trend(s) you might expect between these two variables.

Optimal diet models are largely derivatives of the disk equation Holling (1959), which gives a technique for calculating the rate of return from foraging behavior and takes into account the real-life variables of rate of reward collection, search time, and handling time. What we like best about these models is that you can actually measure the components with real animals and test the theoretical predictions.

Holling actually had blindfolded students collect sandpaper disks that he arrayed on a table. He timed their movements and analyzed their strategies to generate this equation that describes the behavior of a forager. Because Holling used an empirical approach he did not actually derive the equation; rather, he fit an equation to his data. This is the equation:

$$N_a = \frac{a'TNP}{1 + a'T_hN}$$

where: N_a = number of prey attacked, a' = the attack rate, T = time, N = prey density, and T_h = handling time. In theory, a forager finds the behavior that maximizes its return under this model. Basically, the forager needs to assess/remember the four elements of this equation. One question to ponder: Can foragers actually do this math?

Theoretical inquiry on the relationship between prey abundance and consumption of prey by predators generates three model predictions

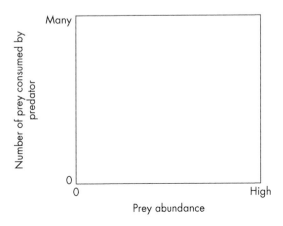

Figure 5.1 Use this empty graph to see if you can predict the relationship between prey abundance and success of a predator finding prey.

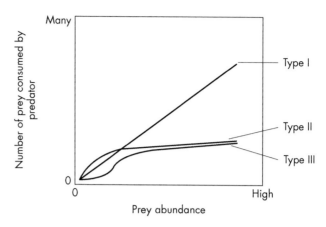

Figure 5.2 The three possible functional responses in Holling's disk experiment.

derived from Holling's equation (Figure 5.2). The first model (Type I) predicts a linear response between prey abundance and consumption by predators. In this model, as the prey reach higher numbers, the predator consumes increasingly more of the prey at a constant rate. The second model (Type II) predicts an initial increase in the frequency of consumption as prey abundance increases; however, the consumption rate reaches a plateau, indicating that beyond a certain prey abundance, consumption rate remains constant. The third model (Type III) differs slightly from this by having little or no increase in consumption when prey are at low abundance. However, as in the second model, after prey abundance passes a certain threshold, consumption rate increases rapidly, followed by a plateau in consumption rate. These three models are depicted in Figure 5.2. Does the trend you drew above reflect any of these predictions? If so, which one?

A Goals

- To gain a first-hand experience of the relationship between prey abundance and foraging rate
- To generate a functional response curve
- To develop an understanding of how different foraging strategies might result in different functional responses

B Questions and hypotheses

Students should consider how different types of foraging strategies might result in different functional response curves. For example, how might the functional response curve of a sit-and-wait predator look? An active-search predator?

Students will perform the role of predator in this exercise, preying upon food items of varying abundance. Which model, Type I, II, or III (or none!) do you think your class data will approximate? Why?

C Methods: Jellybean predators

In this exercise, each participant will act as a foraging animal in patches of varying quality. By recording the rate of prey capture and the prey abundance of the patches, the class will generate enough data to create a functional response figure similar to the above. Each habitat patch is represented by a numbered paper lunchbag filled about 1/3 full with dry lima beans. Mixed into these lima beans are the prey – jellybeans! Dried beans are not considered a food item; each dried lima bean found in the cup will result in the loss of one food item from your total. You will use a plastic spoon as your foraging appendage.

The instructor will keep time for each foraging bout (about 60–90 seconds is usually sufficient for each round). When the instructor says "Go", each student should open his/her paper lunchbag and forage for jellybeans. Students may only use their plastic spoon in one hand to search for and capture jellybeans from the bag, placing the collected jellybeans into a small cup (do not eat the jellybeans...yet!). Students should accumulate as many jellybeans as possible until the instructor ends the round. Each student will record the number on the outside of his/her bag, and the total number of prey items captured from that bag in that round. Then, he/she should return the jellybeans to the bag from which they were taken, mix the beans in the bag a little, and move to a different bag. Repeat this procedure until each bag has been utilized five or more times.

D Interpretation

Worksheet 5.2. As a class, discuss how to set up an appropriate Excel spreadsheet for these data. The spreadsheet should have columns that indicate (i) the identity of the bag, (ii) the total prey abundance of that bag, and (iii) the number of prey captured from that bag each time it was utilized (Figure 5.3). Next, the instructor will collect all the data from each student and assemble a class dataset which will be disseminated to each student. Using statistical software (or Excel), calculate the mean

Bag #	Prey abundance	Prey captured
1	5	3
1	5	2
1	5	3
1	5	4
1	5	3
2	10	5
2	10	7
2	10	8
2	10	6
2	10	7
3	15	11
3	15	12
etc.	...	

Figure 5.3 Sample data sheet.

and standard deviation (SD) for the number of prey captured from each bag. Plot the mean ± SD of jellybeans captured for each different level of prey abundance (use the same axes as the graphs above).

What sort of functional response exists between predator and prey in this interaction? Why? Does this support your hypothesis?

Consider the following examples. Name and justify the type of functional response you might expect to observe between the consumer and its diet.

1. Lions that prey upon mammalian herbivores such as gazelles and zebras
2. A web-weaving spider feeding on flying insects that become tangled in its web
3. A predator that forms a search-image of its prey

III Optimal foraging for jellybeans

Imagine an afternoon spent picking peaches in your local orchard. How might you go about your task? Would you collect every peach from the first tree you encounter before moving to the next tree? Or, would you pick one or two peaches from a tree, pick up your basket, move to the next tree, and repeat?

The answer is probably neither. Picking every peach from a tree would require an exhaustive search of every branch, plus you would probably

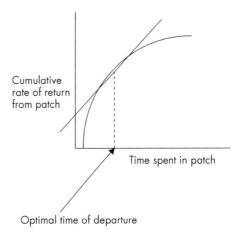

Cumulative
rate of return
from patch

Time spent in patch

Optimal time of departure

Figure 5.4 A graphical representation of the marginal value theorem. The curve represents the yield rate for the environment, not for a given patch. As the cumulative rate of return begins to drop off, the animal perceives that its food yield/unit time is decreasing. The tangent drawn to the curve intersects the curve at the point when the animal is predicted to leave the patch.

need a ladder to climb the tree. It hardly seems worth the effort given there are additional trees! On the other hand, picking just a couple of peaches before moving to the next tree means you will be doing much more walking than necessary; a few more minutes spent at each tree would yield many more easily accessible peaches with much less transit time. The solution you would probably employ would be to pick peaches from a tree for a while until you perceive that the peaches are getting more difficult to reach; it's taking you more and more time per peach. Then, you would pick up your basket and move to the next tree and pick from that tree until you reach the same conclusion. Without realizing it, you are solving the same problem many foraging animals encounter: "How long should I stay in this resource patch before moving to the next patch?"

The Marginal Value Theorem (MVT) is the scientific expression of this question (Figure 5.4). Charnov (1976) first articulated the MVT. He postulated that animals should use the information at hand to predict the future value of a resource patch and make decisions about patch departure based on their assessment of that value. If an animal is foraging optimally, it should leave a resource patch when its current rate of return falls below the average rate of return for all patches (i.e. the whole environment). What information must an animal have in order to be an optimal forager according to the MVT? First, the animal must be able to assess its rate of resource intake. Second, it must have some knowledge – either learned or innate – about the average rate of return for the whole environment. Furthermore, the animal must be able to compare these two rates and decide whether to move or not.

The MVT works if the forager can instantaneously assess the value of the patch. Think about a sit-and-wait predator, such as a snake, which rejects animals that are too large or too small as they walk by, waiting for one that is large enough to be worth the effort, but not so large that it can't be handled. A kind of reverse logic can be applied here to assess foraging yields for animals. Under the MVT, food should be left in a patch after an animal quits the patch, as departure is determined by the relationship between the yield rate for that patch and the yield rate of the larger environment, rather than simple complete depletion of the food resource.

Behavioral ecologists often refer the MVT while studying foraging behavior. By operating within the same context, and collecting similar types of data, ecologists can compare the foraging behavior among different types of animals. Common types of data collected include: (1) patch quality – the relative density of prey items in a patch; (2) residence time – this is the total time the focal animal spends in a particular patch (or the total time you spend at one peach tree); (3) giving up time (GUT) – this is the time between the last food item obtained and the animal's departure from the patch (or the time between when you pick the final peach from one tree and begin to move to the next tree) (McNair, 1982); (4) giving up density (GUD) – the density of food in the patch when the animal leaves the patch; (5) search time – the time it takes for an animal to find its next food item; (6) handling time – the time it takes to capture, subdue, and consume a particular food item (this image may be comical for a herbivore munching grass or our peach-picking analogy, but consider a cheetah stalking patchily distributed gazelles); and (7) transit time – the time it takes to travel from one patch to a new patch.

The GUD, the density of food remaining in the patch, is, in fact, often measurable. If you know the GUD then you can extrapolate the yield rate of food of the area (Brown et al., 1997). Brown et al. (1997) show that while gerbils and larks appear to coexist on the same seed resources in the Negev desert, gerbils have lower GUD (less food left behind) under virtually all conditions (bush versus open habitat, stabilized versus semi-stabilized sand habitats), indicating that gerbils are able to more efficiently exploit the food resource. The authors postulate that larks can coexist by being "cream skimmers", taking food from only high-yield patches in a landscape with high spatial and temporal variability in seed abundances. Larks may rely on insects, fruits, or smaller seeds. Or, larks may rely on adjacent rocky habitats. In contrast to the cream skimming larks, gerbils are "crumb-pickers". Larks do not deplete the food patches, from the gerbils' perspective, but gerbils make the patches unacceptable to larks. Larks must, therefore, have ways of finding patches more quickly than gerbils (see Lab 3 for more thoughts on the importance of quick discovery of food) or exploit other food resources.

If patch assessment is a continuous process while an animal is in a patch, the GUT models are more appropriate. In GUT models, residency in the patch increases with the quality of patch. These models fit well for nectivores and can be used to analyze the rewards that plant present in order to secure pollinator visits (Breed et al., 1996).

How might the environment affect an animal's foraging decisions? For example, the dispersal of patches may significantly affect the decision an animal makes. Let's return to our orchard; how might your decision be influenced if the next peach tree were 500 m away, rather than 5 m away? Suddenly, those peaches you have to stand on your tiptoes to reach look a lot more appealing. Add to this the low, but realistic, chance that you could be eaten by a lion walking between trees, and you gain an appreciation for the mental and/or evolutionary calculus behind the costs and benefits associated with the decision to stay or to move to another resource patch.

In this exercise, you play the role of the foraging animal, your prey are jellybeans and M&Ms, the foraging patches consist of trays or paper sacks of dried beans, and the whole environment is your classroom and all the bags of beans.

A Questions and hypotheses

1. If foragers are able to employ the MVT, how frequently would you expect them to visit patches of high, medium, and low quality?
2. If foragers are not able to employ the MVT, how frequently would you expect them to visit patches of high, medium, and low quality?
3. If foragers are able to employ the MVT, how long would you expect them to spend in high value patches compared to low value patches?
4. If foragers are not able to employ the MVT, how long would you expect them to spend in high value patches compared to low value patches?

B Methods

Divide into groups of four people and designate two foragers and two data recorders. The resource patches are bags/trays filled with dried beans, jellybeans, and M&Ms. Jellybeans and M&Ms are the food items; dried beans are not considered a food item. Each team has a cup that represents the nest and each forager has a plastic fork that represents his or her prey-capturing adaptations. Complete prey handling consists of digging through a patch with the fork, finding an edible item, capturing it with the fork, and transporting it back to the cup using only the fork to place it in the cup. Foragers will forage for an unknown amount of time, to be determined by the instructor. When the instructor calls time, only the jellybeans and M&Ms in the cup count towards the team's prey total. One point is subtracted for each lima bean found in the cup, so forage carefully!

Data recorders follow the activities of the foragers and record the following data: bag number, amount of time spent at each bag by their forager, number of each prey type collected at each bag, and total length of each foraging round.

At the end of a round, each team counts its prey totals and records the data. Team members then trade duties and the game is repeated.

Rules

- The forager can lift and carry as many prey items on the fork as he or she can manage, but cannot hold the items on the fork using his or her other hand.
- Items on a forager's fork are off-limits to other foragers, and students should not intentionally cause another forager to drop his or her jellybeans.

C Interpretation

What is the correlation between patch value and time at patch (Worksheet 5.4)? See Appendix II for instructions on how to calculate a correlation in Excel if you are unfamiliar with this procedure. Based on your calculation, prepare to discuss in class whether what you found agrees with the trend that you predicted based on the graph? Do the foraging humans appear to be operating by the MVT? Why or why not?

References and suggested reading

Blumstein, D. T., Mari, M., Daniel, J. C., Ardron, J. G., Griffin, A. S., and Evans, C. S. (2002). Olfactory predator recognition: Wallabies may have to learn to be wary. *Anim. Cons.* **5**, 87–93.

Breed, M. D., Bowden, R. M.,Garry, M. F., and Weicker, A. L. (1996). Giving-up time variation in response to differences in nectar volume and concentration in the giant tropical ant, *Paraponera clavata. J. Ins. Behav.* **9**, 659–672.

Brown, J. S., Kotler, B. P., and Mitchell, W. A. (1997). Competition between birds and mammals: a comparison of giving-up densities between crested larks and gerbils. *Evol. Ecol.* **11**, 757–771.

Charnov, E. L. (1976). Optimal foraging, the marginal value theorem. *Theor. Pop. Biol.* **9**, 129–136.

Holling, C. S. (1959). Some characteristics of simple types of predation and parasitism. *Can. Entomol.* **91**, 385–398.

Kamil, A. C., Krebs, J. R., and Pulliam, H. R. (Eds.) (1987). "Foraging Behavior." Plenum Press, New York.

Krebs, J. R. and Davies, N. B. (Eds.) (1997). "Behavioural Ecology: An Evolutionary Approach," 4th Ed. Blackwell Science, Oxford, Malden, MA.

Krebs, J. R. and Kacelnik, A. (1991). Decision-making. In "Behavioural Ecology: An Evolutionary Approach" (Krebs, J.R. and Davies, N.B., Eds.). Blackwell Scientific Publications, Oxford, UK.

McNair, J. N. (1982). Optimal giving-up times and the marginal value theorem. *Am. Nat.* **119**, 511–529.

Miller, L. E. (Ed.) (2002). "Eat or Be Eaten: Predator Sensitive Foraging Among Primates." Cambridge University Press, UK.

Pusenius, J. and Ostfeld, R. S. (2000). Effects of stoat's presence and auditory cues indicating its presence on tree seedling predation by meadow voles. *Oikos* **91**, 123–130.

Rothley, K. D., Schmitz, O. J., and Cohon. J. L. (1997). Foraging to balance conflicting demands: Novel insights from grasshoppers under predation risk. *Behav. Ecol.* **8**, 551–559.

Stephens, D. W. and Krebs, J. R. (1986). "Foraging Theory." Princeton University Press, Princeton, NJ.

They were too slow for their own good.

Worksheet 5.1
Herbivore–predator interactions in a system with three trophic levels.

Which herbivore strategies were tested?

Which worked the best?

Which predator strategies were tested?

Which worked the best?

If you got to play both roles, which did you prefer, predator or herbivore? What is it like to find yourself in each of these roles? What was your state of mind as you played each one?

Why did the best strategies work better than others?

For each strategy you tested, provide an analogous example of that strategy in animal behavior.

Worksheet 5.2 The disc equation and decision rules in foraging.

Your datasheet	
Bag #	Prey captured

Worksheet 5.3 Optimal foraging for jellybeans.

Bag Number:	Time Spent (sec)	Prey		Bag Number:	Time Spent (sec)	Prey	
		Jellybeans (#)	M&Ms (#)			Jellybeans (#)	M&Ms (#)

Worksheet 5.4 Marginal value theorem calculations.

1. Calculate the total return for your team's foragers. Jellybeans = 1 point, M&Ms = 5 points. What was your total return?_____
   How did your group do compared with other groups?

2. Calculate the total rate of return for your foragers.
   Total points from above/total time foraging: _____

3. Calculate the average time spent per bag.
   Number of bags visited/total time foraging: _____

4. Using Excel or by hand, calculate the mean number of trips for low, medium, and high value patches. (Your instructor can give you the value of each bag you visited.) See Appendix II for help using Excel.

5. Following the model in Appendix II, enter your data into Excel using Patch #, patch value, and time at patch as your variables. Graph the relationship between patch value and time at patch. When you have a version of the graph that you're happy with, print it out and attach it to this worksheet.

6. Looking at your graph, can you detect a trend in your data? What is the trend, if present?

7. Following the model in Appendix II, perform a correlation analysis on the relationship between patch value and time at patch. What is the correlation between these two variables? Does this agree with the trend you suspected from the graph?

8. In this exercise, do the foraging humans appear to be operating according to the MVT? Why or why not?

Chapter 6
Navigation

How does an animal find its way from home to food, or back to home after finding food? This is a surprisingly complex question, and the answers depend on learning and memory, its sensory abilities, and genetically stored information.

Navigation is the ability of an animal to move in a planned direction, to travel between two predetermined points, or to return to a central location, such as a nest, after leaving for activities such as foraging and mating (Frankel and Gunn, 1961; Bell, 1991; Wiltschko and Wiltschko, 2003). Here, we're primarily interested in movements that require sensing the location of an important environmental stimulus or location and moving directionally so that the animal finds its correct location. In this lab, we investigate three progressively more complex modes of animal navigation.[1] The first is best done indoors, the second could be done either indoors or outdoors, and the third is a field problem. You will not have time to pursue all three topics in a single laboratory; your instructor will probably choose one topic for the laboratory, divide the students into groups to pursue different topics, or use more than one laboratory period. Sections of the lab not covered by your class group would be excellent beginning points for individual projects.

[1]Very simple movements in response to environmental stimuli, called kineses, are covered in Lab 11, Habitat Preferences and Choice Behavior.

1 Counterturning and course maintenance

The most basic navigational strategies in this lab involve the maintenance of a consistent course, or direction of travel. If an animal's course is a straight line, then barriers to its movement, such as rocks, trees, or rivers, may cause it to temporarily change course. To re-establish the original direction of travel, the animal must "counterturn" (in some studies, this is called "turn alternation"). This means making a turn in the opposite direction of the turn that was forced by the obstacle. In this classic test for counterturning (Figure 6.1), a test animal is forced to establish a movement direction, running or walking down the arm of a maze. It then reaches a point where it must turn. After following the second branch of the maze, it reaches a decision point at which, it can make a counterturn which re-establishes its original course, or make a reverse turn. In an experimental context, a number of animals can be asked to move through the maze, and the frequency of counterturns then can be compared with random (50:50) turning. The distance of the second arm of the maze and the time taken to traverse it affects the likelihood of counterturning; as distance and time increase, the animal "forgets" the needed course correction and is more likely to make a random choice.

Early work on counterturning in insects (Dingle, 1961, 1964a, b, 1965) set the stage for consideration of counterturning as a basic behavioral mechanism in animal navigation (Kupferman, 1966; Hughes, 1967; Beale and Webster, 1971). Most of the recent literature that highlights counterturning as an important mechanism deals with how moths find odors (e.g. Vickers, 2000), but most animals probably display counterturning, if they are given the opportunity. Zeil (1998) found, for example, that fiddler crabs counterturn to maintain course constancy when their path is disrupted by moving a turntable on which the crab is walking.

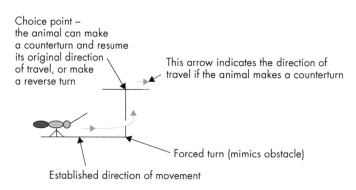

Choice point – the animal can make a counterturn and resume its original direction of travel, or make a reverse turn

This arrow indicates the direction of travel if the animal makes a counterturn

Forced turn (mimics obstacle)

Established direction of movement

Figure 6.1 Apparatus for studying counterturning.

Clearly, learning and memory play an important role in navigation. Animals must be able to remember their travel path, either by storing information such as the angles of their turns and the distances they go, or the direction of travel relative to fixed landmarks, or beginning points. How might the length of the second arm of the maze and the time taken to traverse it affect the likelihood of counterturning?

A Goals

- To learn about counterturning as a basic example of animal navigation
- To test hypotheses about counterturning in a laboratory setting

B Methods

In this lab, you will be provided with either sow bugs or milkweed bugs, a 20-cm square piece of Styrofoam or cardboard to use as a base, strips of 5-mm wide balsa, pins to fix the balsa to the base, and a small light to serve as an aversive stimulus. Start out by constructing a simple maze, following the pattern in Figure 6.1. The initial strip should be 5 cm long, and the second strip (following the forced turn), 1 cm. Set the base in a cardboard box so that your movements, or movements of classmates, are less likely to affect the outcome of the experiment. Start a sow bug or milkweed bug at the beginning of the maze, and shine the light so that it stimulates the organism to move toward the forced turn. For milkweed bugs, a small paintbrush is useful for moving them and herding them to the right spot. Gentle use of forceps works well on sow bugs. *Once the animal has completed the forced turn, turn off the light*; the sow bug or milkweed bug should continue to the choice point, but now won't be influenced by the light. Also be careful to position yourself behind the animal's direction of movement when observing it after the forced turn, so you won't influence its turn choice. You might try initial sample sizes of 10 animals per test, but you may need to increase that number to obtain convincing results.

C Questions and hypotheses

Once you've worked out the technique, try testing these hypotheses:

1. The probability of counterturning depends on the length of the initial leg of the maze. What would you predict here? Would you think it will take

some travel to learn the original direction, or is travel over a short distance adequate? You can test this by varying the length of the segments of the apparatus.

2. The probability of counterturning is dependent on the length of the second segment of the maze. Does the animal "forget" the direction of the forced turn? If so, is it really forgetting, or does the relevance of the forced turn simply diminish with distance traveled? How would you design an experiment to test this?

3. If animals have a turning preference (tend to turn either right or left most of the time) this will affect your analysis. Test a number of animals, maybe 10, on a T-shaped maze (testing each animal 10 times will give an adequate sample size) and count the number of right and left turns made by each. Looking at each bug separately do any of your individual animals display a turning preference? If so, are all of them the same, or do you have a mixture of right- and left-handed animals? How would you use this information as a control for hypotheses 1 and 2?

4. Do humans counterturn? Try the experiment on blindfolded humans. You can use essentially the same experimental design, but you'll obviously need a larger arena for the behavior.

5. If you have time, you can try interesting variations on the counter-turning exercise. In these you allow the animal to walk through the first segment, the forced turn, the second segment, then the animal could encounter:
   - A wide-open plane in which it could go any trajectory ranging from +90 to −90 degrees. The angle the animal walks is recorded. Create and test a hypothesis for how the length of the second segment of the maze would affect the angle of movement.
   - A choice of a 90 degree left-hand turn, a straight ahead path, or a 90 degree right-hand turn. Theoretically, many will counterturn, some will go straight, few/none will reverse turn.

D Interpretation

Is counterturning, which seems fairly artificial when you have animals walking on mazes, an important mechanism in the behavior of animals in the field? Did your bug(s) counterturn consistently? Was the probability of counterturning affected by the length of the first or second segments of the maze? Do you think there are other factors that affect the probability of counterturning that were not tested in this experiment?

Writing an abstract (a short summary) of the experiments you conducted today may help you to organize your thoughts. Your abstract should include a brief introduction to the topic, the hypotheses you were testing, a brief description of the experimental setup, and your findings and conclusions. It should be no more than 500 words.

Your instructor may wish to have the class discuss this question after you've completed the experiments. If you spend considerable time on exploring several of the questions and hypotheses concerning counterturning, your instructor may ask you to write a lab report on this topic.

II Path Integration

Imagine an animal that has a nest or a den, and that forages over a fairly large range. While it is foraging it may make many meandering turns, following cues to food locations. When it is time for the animal to return to its nest or den, it would be very inefficient for it to retrace its steps. (This would be sort of like repeated counterturning, at each turnpoint going the opposite direction than the outbound pathway.) Many animals, including ants (Knaden and Wehner, 2006), rats (Benhamou, 1997), and dogs (Seguinot et al., 1998) can calculate the direct path back home. This is called "path integration" (Figure 6.2). To accomplish path integration, the animal simply computes its return distance and direction vector from the vectors joining the locations on its route. This saves retracing steps. While a variety of animals have been shown to be able to path integrate reasonably accurately in the absence of landmarks or other cues (Biegler, 2000; Hagvar, 2000), long-distance orientation using path integration usually incorporates corrections from landmarks or celestial cues.

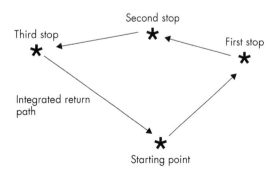

Figure 6.2 The route used in a simple test of path integration. If the animal is able to path integrate, it uses the integrated return path to the starting point. If it cannot path integrate, it may retrace its route back through the second and first stops, then to the start.

A   Goals

- To learn about path integration as a basic example of animal navigation
- To test the abilities of humans to use path integration
- To test hypotheses about path integration in a laboratory setting

B   Method

Experiments with path integration may be easily done with dogs or humans. In this lab, we will use humans. Blindfold your subject (a willing labmate will do), and lead them from the starting point to a series of stops (the route could include two, three, or more). Then, ask them to navigate, still blindfolded, back to the original location. Once they think they've arrived at the starting point, record the distance between their location and the starting point, and the angular deviation between the line they needed to travel from the last stop to reach the starting point, and the actual line they took.

C   Questions and hypotheses

1. Does the performance of your subject improve if you have them do repeated tries using the same starting and stopping points? Do they improve if you have them do a series of trials with different stopping points for each?
2. Does the number of stops affect the performance?
3. Does the distance traveled affect the performance? In other words, if the distances between stops is 3 m, do you get a different result than if its 10 m?

D   Integration

Use Worksheet 6.1 to organize your information. Can you think of other interesting questions that can be answered using this simple experimental paradigm? When humans are not blindfolded they generally use landmarks for orientation, rather than path integration. What are the pros and cons of path integration and landmark orientation, and what ecological contexts might favor one method of navigation over the other?

III Navigation using polarized light

In the previous sections, we explored simple navigation where movement is oriented with respect to a stimulus, and path integration in which an animal must find the shortest way home after following a complex route. In this section, we will explore the mechanisms involved in navigation over long distances or in complex environments. These new challenges require that animals have two navigational tools: a directional sense or a compass, and an internal representation of their environment – a map. To understand the importance of both senses working together, consider being dropped into the forest with a map only. If you do not know which direction is north, the map is of little use. Conversely, if you have only a compass, but no idea where you are in reference to your destination, knowing the direction of north is of little value. An animal uses its compass and map senses together in order to accomplish impressive navigational feats including complex foraging trips, and long distance migration.

What environmental cues do animals use to form a compass and a map? Complex spatial orientation requires the acquisition and processing of environmental cues that allow an animal to plot a course (Collett et al., 1999; Wiltschko and Wiltschko, 1999). Animals may use a variety of sensory modalities to accomplish navigation; some of these are familiar to us, and some are not. The stimuli animals can use include odors, sound, geomagnetism, wave action, landmarks, stars, and the sun. Animals often have the ability to use a combination of several of these. In today's exercise, we will navigate using an environmental cue with which humans are wholly unfamiliar, polarized light from the sun (Wehner, 1976; Wehner and Rossel, 1983; Goddard and Forward, 1991; Lambrinos et al., 1997).

A Goals

- To learn what polarized light is, and how it might be used in navigation
- To find out how a compass and map can work together for successful navigation

B Background

The sun radiates non-polarized light; that is, the photons emitted by the sun vibrate in every possible plane. As these photons pass through Earth's atmosphere, atmospheric molecules block certain planes of vibration.

This creates a pattern of polarization in the sky, with different regions of the sky allowing different planes of polarized light to pass through to the ground.

We can create a small-scale version of this using an overhead projector and sheets of polarizing film. Polarizing film is essentially a series of narrow, parallel, black bars, spaced very closely together. Light from the projector is non-polarized. By placing polarizing film on the projector, we only allow photons vibrating in the same plane, or axis, as the parallel lines of the filter to pass. This has a dimming effect, much like light passing through tinted glass.

Now that we have created a field of polarized light, what will happen if we place another film on top of the first film, with the axis of the film in the same direction? The answer – nothing! Why? The first film has filtered all photons except those in the same plane as the film. Placing another film in the same orientation does not block any additional photons.

What will happen if we place the second film with its axis perpendicular to the first film? The answer – completely black. Why? The second film blocks 100% of the remaining photons that the first film allows to pass.

The light we see in the clear sky is polarized, but we can't detect the polarization. By looking through a polarizing film at the sky, we can find regions of the sky where the light is polarized in parallel with the orientation of our filter's axis, and we can find regions of the sky where light is polarized perpendicular to the axis of our filter (Figure 6.3). In other words, by looking

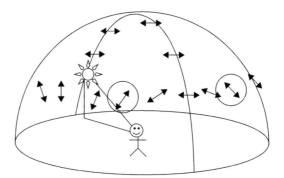

Figure 6.3 Polarized light as seen by an observer at the center of the diagram, looking up at the sky. In this drawing, the sky is represented as a half sphere, with a cartoon of a human looking up at the inside of that sphere. The line that goes from horizon to horizon, and passes through the sun is the solar meridian. Light along the solar meridian (the line from the sun through the zenith) is scattered parallel to the horizon. To either side of the meridian, light is scattered in all directions. The arrows indicate the angle of polarization of the light on the inner surface of the sphere. Imagine that you are looking up into the sky; using a polarizing filter, you would be able to detect these angles. Each direction of polarization is mirrored on either side of the zenith; therefore, at any given time there are two spots in the sky with mirror-image polarization.

through this film, we can see light and dark areas in the sky. Think of the sky as a hemispheric dome with the observer at the center. Great circles drawn perpendicular to the axes of polarization of any two points in the sky will intersect at the sun. Thus, even if clouds cover the sun, its position can be discerned as long as there are two clearings in the clouds. We will use polarized light as our compass guide in today's exercise.

C Questions and hypotheses

In today's lab, we will try to navigate with as little errors as possible using polarized light. We will collect data to determine if the length of the path and the number of turns are related to the error in navigation. What sort of relationship might you predict between these variables?

Box 6.1   How do animals perceive polarized light?

The ability to use polarized light is best understood in insects (Horvath and Wehner, 1999). The compound eye of insects comprises thousands of individual photoreceptors called ommatidia (Figure 6.4). Each ommatidium is itself made up of nine visual receptor cells. Each visual cell contains the photopigment rhodopsin, a dipolar molecule. Rhodopsin maximally absorbs photons whose vibrational axis is parallel to rhodopsin's axis. In insect eyes, the rhodopsin molecules are all aligned parallel to each other inside the microvilli of the ommatidia. The parallel orientation of the rhodopsin molecules makes the ommatidia sensitive to polarized photons vibrating in the same plane as the rhodopsin.

The ommatidia sensitive to polarized light are concentrated on the upper front region of the compound eye. When the insect is horizontal to the ground, this points the polarized light-sensing ommatidia towards the sky. The insect can simultaneously observe two points in the sky, one from each eye, and use the difference in polarization between the two for orientation.

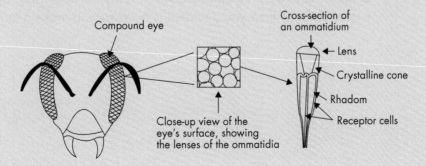

Compound eye

Close-up view of the
eye's surface, showing
the lenses of the ommatidia

Cross-section of
an ommatidium

Lens

Crystalline cone

Rhadom

Receptor cells

Figure 6.4   Schematic drawing of a bee's eye.

D Methods

Divide into groups of four people; each group will be assigned a number, 1 through *n*. Each group will then divide into two-person teams – A and B. The instructor will have hung flags around campus labeled for each group and team (e.g. "1A" is group 1, team A). Each two-person team will be given a small map with the location of their flag marked on it. Do not allow the other two-person team in your group to see this map! Your job is to create a travel log to the flag that will serve as a navigation map using polarized light as your compass.

Hold the polarizing film about 45 degrees above the horizon. Rotate yourself and watch as the sky gets darker and lighter as you turn. Agree as a group concerning which point (perhaps near the horizon above a recognizable landmark) in the sky, you will designate as 0 degree.

1. Team A is given the map for flag A. They create a travel log such as this (Table 6.1):

Table 6.1 Sample travel log.

Leg	Degrees	Distance (Paces)
1	180	75
2	90	150
3	45	40
4	100	35

As they navigate to their flag Team B does the same thing for Flag B.

The path can be represented by a travel log like Figure 6.5. Measure the distance in paces. It might be easier to stay on sidewalks, but this is not essential. Each time you decide it is necessary to make a turn, stop, find

Figure 6.5 Sample travel log.

0 degree in the sky again, and decide which direction your new trajectory is with reference to 0 degree. This diagram is for illustrative purposes; do NOT draw a map for your group members. They must try to navigate by your travel log alone.

2. After creating a travel log to the flag (Worksheet 6.2), remove the flag and return to the common start point. Each team within a group exchanges flags and travel logs, but NOT initial maps; return these to the instructor. Now, team A must try to hang the flag from team B back in its original location, using the travel log as a map, and the polarized light of the sun as a compass. Team B must do the same for flag A. After every team has re-hung the flags according to their directions, measure the distance between where the flag *is* and where it *should be*! The group with the smallest total displacement wins!

E   Interpretation

What natural phenomena might make the use of polarized light unreliable? Humans often use sun-compass orientation, with the sun or a magnet as a referent for determining north, south, east, and west. Based on your experience with your ordinary mode of orientation using compass directions and, in this lab, using polarized light, do you think polarized light adds to the accuracy of an animal's orientation capabilities?

Today's exercise gives us the opportunity to perform a simple statistical test – a correlation (see Appendix II for details on how to calculate a correlation using Excel). This may be especially valuable for students who intend to do their own field project. With a correlation, we can determine whether there is a linear relationship between the flag displacement, and the length and complexity of the path. Correlations tell us if there is a positive or negative relationship between two measured variables; in other words, as one variable increases or decreases, what happens to the other variable? It is important to remember that just because two variables are correlated, that doesn't mean that one variable is the actual cause of the change in the other variable.

Now, make a graph showing the statistics you've just done. You can use the Chart Wizard to walk through the steps of making a graph, or follow the detailed instructions in Appendix II. Make sure that your graph has a title and axis labels that make it very clear what the graph is showing (it's OK to have a title that is more like a caption, with several sentences, if necessary). You will also want to include your *r*-value as a text box somewhere on the graph. Your instructor may ask you to print out your graph, and hand it in.

References and suggested reading

Beale, I. L. and Webster, D. M. (1971). The reference of leg movement cues to turn alteration in woodlice (*Porcellio scaber*). *Anim. Behav.* **19**, 53–56.

Bell, W. J. (1991). "Searching Behaviour: The Behavioural Ecology of Finding Resources." Chapman and Hall, London.

Benhamou, S. (1997). Path integration by swimming rats. *Anim. Behav.* **54**, 321–327.

Biegler, R. (2000). Possible uses of path integration in animal navigation. *Anim. Learn. Behav.* **28**, 257–277.

Collett, M., Collett, T. S., and Wehner, R. (1999). Calibration of vector navigation in desert ants. *Curr. Biol.* **9**, 1031–1034.

Dacke, M., Nilsson, D. E., Warrant, E. J., Blest, A. D., Land, M. F., and O'Carroll, D. C. (1999). Built-in polarizers form part of a compass organ in spiders. *Nature* **401**, 470–473.

Dingle, H. (1961). Correcting behavior in boxelder bugs. *Ecology* **42**, 207–211.

Dingle, H. (1964a). Correcting behaviour in mealworms (*Tenebrio*) and the rejection of a previous hypothesis. *Anim. Behav.* **12**, 137–139.

Dingle, H. (1964b). Further observations on correcting behaviour in boxelder bugs. *Anim. Behav.* **12**, 116–124.

Dingle, H. (1965). Turn alternation by bugs on causeways as a delayed compensatory response and the effects of varying visual inputs and length of straight path. *Anim. Behav.* **13**, 171–177.

Etienne, A. S., Maurer, R., Berlie, J., Reverdin, B., Rowe, T., Georgakopoulos, J., and Seguinot, V. (1998). Navigation through vector addition. *Nature* **396**, 161–164.

Frankel, G. S. and Gunn, D. L. (1961). "The Orientation of Animals." Dover, New York.

Goddard, S. M. and Forward, R. B. (1991). The role of the underwater polarized-light pattern, in sun compass navigation of the grass shrimp, *Palaemonetes vulgaris. J. Comp. Physiol. A* **169**, 479–491.

Hagvar, S. (2000). Navigation and behaviour of four Collembola species migrating on the snow surface. *Pedobiologia* **44**, 221–233.

Horvath, G. and Varju, D. (1995). Underwater refraction – polarization patterns of skylight perceived by aquatic animals through Snell's window of the plat water-surface. *Vision Res.* **35**, 1651–1666.

Horvath, G. and Wehner, R. (1999). Skylight polarization as perceived by desert ants and measured by video polarimetry. *J. Comp. Physiol. A* **184**, 1–7.

Hughes, R. N. (1967). Turn alternation in woodlice (*Porcellio scaber*). *Anim. Behav.* **15**, 282–286.

Knaden, M. and Wehner, R. (2006). Ant navigation: resetting the path integrator. *J. Exp. Biol.* **209**, 26–31.

Kupferman, L. (1966). Turn alternation in the pill bug (*Armadillium vulgare*). *Anim. Behav.* **14**, 68–72.

Lambrinos, D., Maris, M., Kobayashi, H., Labhart, T., Pfeifer, R., and Wehner, R. (1997). An autonomous agent navigating with a polarized light compass. *Adapt. Behav.* **6**, 131–161.

Lawson, P. A. and Secoy, D. M. (1991). The use of solar cues as migratory orientation guides by the plains garter snake, *Thamnophis radix*. *Can. J. Zool.* **69**, 2700–2702.

Mouritsen, H. (2001). Navigation in birds and other animals. *Image Vision Comput.* **19**, 713–731.

Seguinot, V., Cattet, J., and Benhamou, S. (1998). Path integration in dogs. *Anim. Behav.* **55**, 787–797.

Vickers, N. J. (2000). Mechanisms of animal navigation in odor plumes. *Biol Bull.* **198**, 203–212.

Waldvogel, J. A. and Phillips, J. B. (1991). Olfactory cues perceived at the home loft are not essential for the formation of a navigational map in pigeons. *J. Exper. Biol.* **155**, 643–660.

Wehner, R. (1976). Polarized-light navigation by insects. *Sci. Am.* **235**, 106–115.

Wehner, R. and Rossel, S. (1983). Polarized-light navigation in bees – use of zenith and off-zenith e-vectors. *Experientia* **39**, 642–642.

Wiltschko, R. and Wiltschko, W. (1999). The orientation system of birds – I. Compass mechanisms. *J. Ornith.* **140**, 1–40.

Wiltschko, R. and Wiltschko, W. (2003). Avian navigation: from historical to modern concepts. *Anim. Behav.* **65**, 257–272.

Zeil, J. (1998). Homing in fiddler crabs (*Uca lactea annulipes* and *Uca vomeris*: Ocypodidae). *J. Comp. Physiol. A.* **183**, 367–377.

Worksheet 6.1 Path integration.

For each of the questions addressed in the lab, write down your hypothesis, how many subjects/trials you used to test it, and your results.

1 Hypothesis   #Trials/#Subjects   Results (distance and angle of error)

2 Write a short summary of what you can conclude about path integration in humans, based on your experiments.

3 If you were a test subject, how confident would you feel in your ability to path integrate? What factors (besides removing the blindfold) might help you to feel more confident?

4 What do you think are the limits of humans' ability to path integrate? In what circumstances do we use this ability, and how could we improve it?

Worksheet 6.2 Polarized light orientation.

Your travel log:

Leg	Degrees	Distance (Paces)
1		
2		
3		
4		
5		
6		
7		
8		

Fill out the section below, AFTER completing the lab. You will need to get the data from all of the teams in your lab section.

Team	Accuracy: distance of the flag from the goal	Complexity of route: Number of legs	Complexity of route: distance traveled

Using Excel (Appendix II), calculate the correlations between accuracy and the two measures of complexity – show your result below.

Chapter 7
Territoriality in red-winged blackbirds

Mate choice is one of the most intriguing topics in animal behavior (Creighton, 2001; Wong and Candolin, 2005). In many species, males work very hard to be chosen, establishing territories (Turner and McCarty, 1998), and carrying costly plumage, and/or adornments such as antlers or horns. Females also work very hard to make the correct choice of mate, or to hedge their bets by surreptitiously mating with more than one male (Gray, 1997; Westneat and Mays, 2005). In this lab we'll explore the behavior of territorial male red-winged blackbirds (RWBs) during their mating season.

RWBs, *Agelaius phoeniceus*, are among the most common birds in North America (Figure 7.1). The breeding range of the RWBs extends from northern Canada to Central America; northern populations fly south to overwinter, while southern populations do not migrate. Adult male RWBs are black with striking red epaulets on the wings. Females and juvenile males are not as striking, tending to be mottled brown or gray in color. Males reach sexual maturity after about three years, at which time they begin to engage in interesting breeding behavior.

RWBs are socially polygynous; males establish territories that serve as nesting areas for multiple females (Holm, 1973; Newhouse, 1977). Male territories are located in marshes along the water's edge, typically in strands of *Typha* (cattail) (Ozesmi and Mitsch, 1997). Each male defends the boundaries of his territory by vocalization and visual signals, and by staying in his territory for long periods of time. The males' calls make up a prominent portion of the chorus you hear near marshes

Figure 7.1 The red-winged blackbird (photo by Jeff Mitton).

on a spring evening. Male RWBs are easy to spot as they perch on the tips of tall cattails or other vegetation.

Females choose males with whom to mate based on the perceived quality of each male's territory. Females appear to prefer male territories with more dense strands of *Typha*, where they will build their nests. The nests are shaped like a cup made of marsh grasses and reeds, and are usually attached to live vegetation. Chicks are fed primarily by their mother, and fledgling success is correlated with the number of emerging aquatic insects, such as damselflies and dragonflies, in the vicinity.

A female typically produces two clutches of eggs over the course of one summer. Though she resides in a male's territory, her offsprings frequently are not all sired by that territory's male. By August, territorial behavior of the males diminishes and the final clutch of chicks begins to fledge. The birds fatten up on seeds in large fields, usually mixing with other granivorous birds. They undergo a molt prior to their winter migration, and females usually depart before the males. RWBs overwinter in large mixed flocks until early spring, when the males head back north again to begin establishing territories.

We try to time our observations early in the mating game of RWBs, when males are in the process of establishing and defending their territories. Females are "shopping around" for males with good territories, but probably have not committed to a particular territory yet. We will observe several aspects of RWBs mating behavior, including male site fidelity, male–male territorial interactions, and male–female interactions.

I Goals

- To practice scan and focal sampling in the field (see Appendix I for descriptions of these sampling techniques)
- To observe territorial behavior and mating system of RWBs

II Questions and hypotheses

We'll test three specific hypotheses. None of these, individually, tests for territoriality, but the results of all three, when taken together, will allow you to make a convincing case as to whether or not male RWBs are territorial.

- Our first hypothesis is that male RWBs have a high level of site-fidelity.
- Secondly, we hypothesize that male-calling increases when another male is nearby. This can be separated into an assessment of the visual or vocal presence of another male.
- We'll observe male–male interactions to see if males chase each other from their territories. This will allow us to diagram territorial boundaries.
- Finally, we'll test whether male-calling increases when a female is nearby.

III Method-testing hypothesis 1 (site fidelity)

1. Create a sketch map of a pond and the surrounding area.
2. Choose a male RWB. Without startling the bird, arrange your group around the bird's perch so that you can observe him when he flies. During your observations, write down when your bird calls and any other RWB calls that you can hear.
3. Indicate on your map where your bird is perched.
4. Record the length of time the bird remains at that perch (see data table).
5. When the male leaves the perch, track him to his next perch.
6. Indicate this new perch on your map.
7. Record the length of time the bird remains at that perch.

8. Repeat 5–7 for as long as possible.
9. If your focal male flies out of observational range, record the length of time he is gone from his territory. Keep watch on his previous perches so that you know when he returns.
10. When your male returns, repeat 4–8 for as long as possible.
11. On your map of the pond, sketch what appear to be the boundaries of your male's territory; you can use male chases and occupation of areas as evidence for territorial boundaries.
12. Record your data in Worksheet 7.1.

IV Method-testing hypotheses 2 and 3 (vocalizations)

1. Choose a male redwing, and record the number of calls per minute made by the male.
2. Record the frequency of male-calls when alone, when another male enters his territory (you may simulate this by playing a recorded RWB call), and when a female enters his territory.
3. Record your data in Worksheet 7.2.
4. Given adequate time (we recommend 30 minutes) with one male, move on, and record the activity of another male.

V Interpretation

After you have completed the lab and recorded your data in the Worksheets 7.1 and 7.2, take a moment to consider the interpretative questions at the end of each worksheet and to jot down some ideas about the answers to the questions. Clearly the sample size from any one student's observations will not be adequate for fully testing the hypotheses. Your instructor, either by gathering your class for a discussion, or by having you share copies of your data, will help you to assemble a more complete picture based on the larger sample size collected by your class. When your class discusses the results, focus on the interpretative questions at the end of each worksheet.

References and suggested reading

Creighton, E. (2001). Mate acquisition in the European blackbird and its implications for sexual strategies. *Ethol. Ecol. Evol.* **13**, 247–260.

Gray, E. M. (1997). Do female red-winged blackbirds benefit genetically from seeking extra-pair copulations? *Anim. Behav.* **53**, 605–623.

Holm, C. H. (1973). Breeding sex-ratios, territoriality, and reproductive success in red-winged blackbird (*Agelaius phoeniceus*). *Ecology* **54**, 356–365.

Newhouse, C. (1977). Territoriality in red-winged blackbird. *Am. Biol. Teacher* **39**, 168–170.

Ozesmi, U. and Mitsch, W. J. (1997). A spatial habitat model for the marsh breeding red-winged blackbird (*Agelaius phoeniceus* L) in coastal Lake Erie wetlands. *Ecol. Model.* **101**, 139–152.

Turner, A. M. and McCarty, J. P. (1998). Resource availability, breeding site selection, and reproductive success of red-winged blackbirds. *Oecologia* **113**, 140–146.

Westneat, D. F. and Mays, H. L. (2005). Tests of spatial and temporal factors influencing extra-pair paternity in red-winged blackbirds. *Mol. Ecol.* **14**, 2155–2167.

Wong, B. B. M. and Candolin, U. (2005). How is female mate choice affected by male competition? *Biol. Rev.* **80**, 559–571.

Worksheet 7.1 The site fidelity hypothesis.

Name:_____

Watch a male bird and number each perch it lands on and record the time spent on that perch. Use your numbering system so that when it revisits a perch, that observation is recorded as the same perch number. Use the space to the right of the table to sketch a map of your bird's territory, showing the perches and the spatial relationship among them. In the second table, give the cumulative time in and out of the territory for your bird.

Perch #	Time (min:sec)

Total time: (min:sec) in territory	Total Time: (min:sec) away from territory

Interpretation

After observing a male RWB, describe his territory. How large is it? What sort of vegetation is included in it? How many different perches did he have? Did he seem to have a favorite perch? Do there seem to be any other males with abutting territories? Our hypothesis is that males should stay in their territories for long periods of time. Did your male seem to stay in his territory? What are several reasons why it is a good idea for males to have high site fidelity? Despite our best efforts, the presence of humans can influence the way an animal behaves. How do you think your presence might have affected your focal male?

Worksheet 7.2 Test for the effects of other birds on a focal male's calls.

Calls/minute: while alone	Calls/minute: when male intruder is present	Calls/minute: when female is present

Interpretation

Using an Excel spreadsheet, do a t-test between the calls per minute when a male or a female is present. Following the instructions in Appendix II, enter the data, obtain, and interpret the result of the t-test. Does the mean call rate differ? Why do you think this might be? Do territorial calls to other males sound different than calls that might attempt to attract females?

Male RWBs invest a lot of time and energy into acquiring and defending a territory. As such, we might expect them to behave very defensively when they think their territory is being invaded. How did the males' calls/minute change when they encountered another male (either real or recorded)? How did their call rate change in the presence of a female? Do you suppose the signal means the same thing to both receivers? Why/why not? Did the males do anything other than vocalize in response to intruder males?

Chapter 8
Vigilance and selfish herds

Why would an animal choose to live in a group (Hamilton, 1971; Amano et al., 2006)? One of the big advantages of group-living comes from being able to exploit the observational powers of other members of the group (Bekoff, 1995; Verdolin and Slobodchikoff, 2002). This can work as a protection against predators or in finding food – many eyes or ears or noses are better than those of a single animal. The focus of this lab is on how animals use vigilance as a part of their anti-predator behavior.

Quantifying vigilance provides an easy way to measure anti-predator behavior. An animal is vigilant when it has its head up and is observing the landscape around the group. Published studies of vigilance usually employ a very simple measure – an animal with its head up is deemed vigilant, an animal with its head down (generally feeding) is considered non-vigilant. You can see a dramatic (if trivial) shift from non-vigilance to vigilance by making a loud noise in a restaurant; people's heads pop up and they will visually scan the room in response to the disturbance.

Defined as head-up versus head-down, vigilance can be easily studied in a variety of animals. Simple observations of vigilance can be overlaid with more complex scientific hypotheses. Perhaps the most intriguing of these is whether vigilance is shared evenly among group members. Interesting questions about vigilance are: Do animals in groups gain an advantage from shared vigilance? In a group, are there factors that determine which animal will be vigilant? Is the benefit from vigilance evenly distributed among group members or do some benefit more than others? Do animals that are exposed to human disturbances habituate, making them less vigilant?

Figure 8.1 While both of these bighorn sheep have their heads down, pairing like this allows the animals, jointly, to have a 360 degree view of the landscape (photo by Michael Breed).

Published studies have examined vigilance in geese (Amano et al., 2006), deer (Altendorf et al., 2001), grosbeaks (Bekoff, 1995), prairie dogs (Verdolin and Slobodchikoff, 2002), and domestic cattle (Welp et al., 2004) (Figures 8.1–8.3). These studies all use simple heads-up versus heads-down criteria for defining vigilance, and they demonstrate the variety of interesting hypotheses that can be tested using a very simple methodology.

I Explanations for vigilance

A Selfish herds

Very much related to the question of vigilance is the theory of the selfish herd. Certainly, you would expect close genetic relatives to watch out for one another – natural selection should favor behavior that enhances the survival of close relatives – but why be vigilant for a nonrelative?

Figure 8.2   Typical heads-up posture in a vigilant goose. Note that all the other members of the flock are foraging (photo by Michael Breed).

Figure 8.3   New Forest deer (England) displaying heads-up postures after a disturbance (photo by Michael Breed).

Hamilton's (1971) theory of the selfish herd attempts to explain the existence of herds or flocks of unrelated animals in which vigilance is shared. In a selfish herd, animals mutually benefit from vigilance, but each tries to minimize its own time spent on vigilance and to occupy a location within the herd or flock that is least risky (Clutton-Brock et al., 1999). A simple prediction is that animals should compete for the central, protected, locations and that animals forced to the edge should be most vigilant. The vigilance of these animals helps protect all the animals in the herd or flock, but really they're being vigilant for the purpose of self-protection, not because they're being generous.

Interestingly, the larger the herd or flock, the lower each animal's chances of being preyed upon. This should lead to larger and larger groups. But, counterbalancing this is the fact that competition for food increases as group size increases. Can you think of ways of testing how vigilance, group size, and food availability interact?

B Reciprocal altruism

An alternative thought explains shared vigilance in flocks through reciprocal altruism (Krams et al., 2006). In this case, an unspoken contract exists among the animals, so that each takes its turn being vigilant. The payoff comes from being able to feed while other group members are taking their turn. In a reciprocally altruistic flock or herd, you wouldn't expect competition for location and you might predict that vigilant animals would be more scattered through the group, rather than only the most vulnerable animals being vigilant.

Under reciprocal altruism, vigilance might be shared, over time, among animals that clump together within the herd. So, then you'd predict that vigilant animals would be more randomly distributed in the larger group, but you might be able to spot subgroups that stay together and share vigilance.

C Kin selection

You probably won't have the ability to test a kin selection hypothesis in this lab, but you should consider that a kin selection model would predict that animals live in herds with close relatives, or that they only display vigilance when near close relatives, so that their vigilance benefits relatives, rather than unrelated individuals (Griesser, 2003). In herds of animals like deer or elk, the dominant male in the group may be more vigilant. His vigilance would help him to keep other males away from his herd and also protect his offspring and females with whom he will have offspring in the future from predation.

D Other reasons for flocking

Just to add to the difficulty in thinking about this, consider that herds or flocks might exist for other reasons altogether. Animals may live in groups to take advantage of each other's ability to find food, for example. If you throw a piece of bread to a goose, do the other geese in the flock swarm around that individual and try to steal the food? Of course they do; overall, it may be easier to rely on others to find food for you than to try to always find it on your own.

To develop a hypothesis that distinguishes between the alternative hypotheses of selfish herds and reciprocal altruism, first consider some of the vigilance behaviors your subject animals perform. For example, perhaps your animal holds up its head and scans the horizon. How might this behavior differ if the animals are following a selfish herd model compared to a reciprocal altruism model? For example, can you make predictions about how this behavior might affect the location of those performing the behavior relative to the group?

Clearly to begin to test these hypotheses you need to know how the frequency of performing the behavior varies among individuals. It helps to design your hypothesis in a way that explicitly states the type of numerical data you will need. How would you quantify the position of vigilant individuals? Feel free to incorporate more than one behavior – perhaps vocalization is also part of your focal animal's vigilance behavior. The more testable predictions you can make, the better will be your case for distinguishing among models that try to explain why animals are vigilant.

For background purposes, we present information on a number of examples of species that display vigilance or on situations in which you might observe vigilance. The precise study that you or your instructor will choose will depend on the time of year, your geographic location, and the abundance of possible study animals. Animals which live in groups and which tolerate some exposure to human observers make good choices for the investigation of vigilance. As you read through the rest of this laboratory, think about whether you expect the same evolutionary mechanisms to operate in all species and ecological contexts. If you studied vigilance in a number of species, would the same hypotheses always be supported? You should consider these possible explanations as you construct your hypotheses and design your experiments.

II Background on some potential study species

A Vigilance in flocking ducks or geese

Ducks or geese make great subjects for vigilance studies. Beginning in the early fall and continuing until the beginning of the breeding season, many species of ducks and geese live in flocks. Through much of North America and Canada geese are particularly interesting study-subjects because flocks can be found feeding on land as well as in the water. Can you test hypotheses that relate vigilance to context (foraging in a field versus resting near the water versus at the nest) or group size? An additional factor that affects vigilance is the type of threat the animal or group encounters. You might observe vigilance responses when a human walks near a flock and compare those with what happens when a human with a dog on a leash walks near the same flock. Or you might simulate a hawk by flying a hawk-shaped kite over the geese. Be sure you don't run a-fowl (ha!) of wildlife harassment laws – check with responsible authorities where you are making your observations before doing these types of manipulations and make sure you show respect for the animals. For instance, it would be more acceptable to do the experiment with the dog on a leash, for example, in a park where people often walk their dogs, than in a protected area where dogs are not commonly seen. If you live near the ocean, you can use the same approach with studies of seagulls.

Mallard ducks, which are usually found feeding in water, also display vigilance behavior. Most of the same observations and experiments that we have suggested for geese could be conducted, although you're less likely to be able to make comparisons of land-foraging versus swimming birds. One advantage of mallards is that you can easily distinguish males from females, facilitating tests of hypotheses about gender differences in vigilance.

B Flocking birds at feeders

Bekoff (1995) gives a detailed example of how vigilance might be studied in a flocking bird species at feeders and Cheverton (1996) provides guidelines for student projects of birds at feeders. Our suggestions generally follow those of Cheverton (1996). The best time of year to study flocks of birds at feeders starts in the late fall and continues until the beginning of the nesting season. Once the birds nest the social dynamic among them changes substantially and foraging flocks are much less likely to be observed. Through most of North America the most common birds at feeders in fall and winter will be English sparrows (*Passer domesticus*, sometimes called house sparrows), although in some locations juncos, goldfinches, or house finches will be most common. Other birds at feeders, such as blue jays and cardinals, tend not to flock and are not so amenable for studies of vigilance.

Cheverton (1996) suggests hypothesizing that accessibility of the feeder to cats might affect vigilance. In this case, a feeder that is low to the ground or close to bushes that might hide a cat would be hypothesized to have more vigilant birds than a feeder that is high off the ground and away from places for cats to hide. He also points out that under very cold conditions birds may be closer to starvation and that they may then spend more time feeding and less time being vigilant, even though this would increase the risk of being preyed upon. Another threat is larger birds (falcons, hawks) that would come from above. Does protecting the feeder with a board that blocks potential predators from above change vigilance patterns?

You might also consider how position in the flock affects vigilance. Are the birds nearer the edge (and consequently more vulnerable) more likely to be vigilant or spend more time in each bout of vigilance? If this is the case, can you test a hypothesis that birds compete for central locations in the flock? Is there a male–female difference in the ability to occupy favored positions in the flock? How does nearby movement affect vigilance (you might try rigging a device near the feeder that can be moved by tugging on a string)?

C Deer and domestic livestock

Many of the same lines of reasoning that we've presented for geese and birds at feeders can be applied to herds of deer, cattle, sheep, or horses. How is vigilance shared in these animals and what influences which animals in the group are being vigilant?

D Habituated versus non-habituated wildlife

Habituation in wildlife poses very interesting questions relating to vigilance. Clearly, animals that are around humans much of the time will benefit from ignoring the humans, rather than fleeing every time a human approaches. Animals that needlessly flee will lose considerable time and energy. Can an animal that ignores a human still respond to a real threat, such as a dog or cat? If you pursue this question, there are several informative types of comparisons you can make. You might, for example, compare an urban and a rural population of the same species of animal. Or you could consider comparisons among species in an urban environment. Do some, such as squirrels, habituate, while others, such as sparrows, never habituate? If this hypothesized difference is supported by the data, why might it exist? Are there consequences for the animals if their vigilance is relaxed due to habituation?

III Goals

In today's lab, your instructor will have chosen an animal species that displays vigilance. From this lab, you should learn:

- How to construct ethograms and time budgets
- Differences between scan and focal animal sampling (see Appendix I for a description of these techniques)
- How self-interest and shared interests interact in protecting a group from predators

IV Questions and hypotheses

- Can you design hypotheses that test among the possible explanations for vigilance?
- Do you hypothesize that animals in an urban environment would be more vigilant than their rural counterparts, or less? Why? How would you test this?

V Methods

A. Construct an ethogram for your study species. An ethogram is a list of the possible behavioral acts, with descriptions and definitions for

Table 8.1 Sample ethogram.

Behavior	Number of animals performing behavior at given time
Head up, standing still	
Head up, walking	
Head down, feeding	
Head down, not feeding	
Sleeping	

each act. Having an ethogram allows you to be precise in counting how many acts of each type an animal performs, and it also allows you to accurately communicate your results. Clearly one of the behavioral acts included in the ethogram should be vigilance, which is generally defined as the animal having its head up, and not feeding. See Table 8.1 for a sample ethogram.

In the right-hand column you could record the number of individuals performing each behavior at a given moment. If there are 30 animals in the group, how many are engaged in each activity? A quick digital photograph should help you in making these counts. A snapshot, or instantaneous, count of behaviors in a group of animals is a **scan** sample; before the era of digital photographs, you would have scanned the group and counted the number of animals performing a given behavior. If you have more than one flock or herd, is the distribution of behaviors performed in the ethogram the same, or different for the groups? Does this distribution of behaviors change over time for the group you're watching? If so, why?

B. A time budget is based on the ethogram, but focuses on an individual animal. If you can track individuals for at least short periods of time, you might focus on each animal for a fixed time interval, perhaps five minutes, and then move on to the next animal. This is **focal** animal sampling (see Appendix I). As you make the time budget, also record the animal's position within the group – is it in the center or on the edge?

A time budget looks similar to an ethogram, except that your observations will be entered as amounts of time spent in each activity by an individual animal (Table 8.2).

By constructing time budgets for a number of animals in a group, you can test the hypothesis that behavior is differentially expressed. Remember that the different models for cooperation in vigilance – selfish herd, reciprocal altruism, kin selection – make differing predictions about how animals in the group might divide up the responsibility for

Table 8.2 Sample time budget.

Behavior	Cumulative time spent in activity
Head up, standing still	
Head up, walking	
Head down, feeding	
Head down, not feeding	
Sleeping	

being vigilant. Worksheets 8.1 and 8.2 provide models for recording ethograms and time budgets.

C. Map the spatial occurrence of vigilance within the group. For example, you might predict that in a selfish herd, animals at the edges of the herd would be more vigilant, because they are more at risk to be preyed upon. There would be competition for the more central locations, which would be less risky. Using a grid can help you to assess the spatial distribution of behaviors within a group of animals (Worksheet 8.3).

D. The methods for studying birds at feeders also involve ethograms and time budgets. A feeder stocked with a mix of seed, including corn and sunflower seeds, will attract the widest range of birds. Any commercial or home-made feeder will do; in many localities you'll want to construct the feeder so squirrels won't compete with the birds for seed. You can vary the amount of competition among birds for seed by giving them a small opening, a medium-sized opening, or a large tray from which to feed.

If there are dozens or hundreds of birds at the feeder at a time, you'll need a better method of collecting data than watching and writing down what you see. A still picture taken with a digital camera (even a phone camera will work well for this) will allow you to count the number of birds with their heads up or down at a given moment. Video-recording birds at the feeder allows a more sophisticated analysis, as you can use a stopwatch to time how long each bird spends in vigilance versus feeding. Be mindful, though, that it is easy to record many hours of behavior and much more difficult to analyze those hours of recording.

Once the feeder is established and flocks of birds are visiting, you can settle on one or more species for observation and on experimental hypotheses. Because it is unlikely that you'll be able to identify individual birds, you should think about hypotheses that relate vigilance to environmental influences, or that compare males to females, if they differ so you can identify gender in the field. Obviously, you can also compare among species if you have more than one species at your feeder.

VI Interpretation and points for discussion

At the end of the lab, you should discuss your findings with the other students. These questions should help you get started on the discussion: Based on your observations and data collection, can you make an argument that the animals you have studied are being vigilant for selfish or altruistic reasons? How do you think patterns of vigilance by altruists would differ from vigilance by selfish animals? Does the behavior you have observed seem to fit either pattern?

References and suggested reading

Altendorf, K. B., Laundre, J. W., Gonzalez, C. A. L., and Brown, J. S. (2001). Assessing effects of predation risk on foraging behavior of mule deer. *J. Mammal.* **82**, 430–439.

Amano, T., Ushiyama, K., Fujita, G., and Higuchi, H. (2006). Costs and benefits of flocking in foraging white-fronted geese (*Anser albifrons*): effects of resource depletion. *J. Zool.* **269**, 111–115.

Beauchamp, G. and Ruxton, G. (2003). Changes in vigilance with group size under scramble competition. *Am. Nat.* **161**, 672–675.

Bekoff, M. (1995). Vigilance, flock size, and flock geometry – information gathering by western evening grosbeaks (Aves, Fringillidae). *Ethology* **99**, 150–161.

Cheverton, J. (1996). "Vigilance in feeding birds" http://asab.icapb.ed.ac.uk/practicals/vigilance_feeding_birds.html

Clutton-Brock, T. H., O'Riain, M. J., Brotherton, P. N. M., Gaynor, D., Kansky, R., Griffin, A. S., and Manser, M. (1999). Selfish sentinels in cooperative mammals. *Science* **284**, 1640–1644.

Krams, I., Krama, T., and Igaune, K. (2006). Alarm calls of wintering great tits Parus major: warning of mate, reciprocal altruism or a message to the predator? *J. Avian Biol.* **37**, 131–136.

Griesser, M. (2003). Nepotistic vigilance behavior in Siberian jay parents. *Behav. Ecol.* **14**, 246–250.

Hamilton, W. D. (1971). Geometry for the selfish herd. *J. Theor. Biol.* **31**, 295–311.

Verdolin, J. L. and Slobodchikoff, C. N. (2002). Vigilance and predation risk in Gunnison's prairie dogs (*Cynomys gunnisoni*). *Can. J. Zool.-Rev. Can. Zool.* **80**, 1197–1203.

Welp, T., Rushen, J., Kramer, D. L., Festa-Bianchet, M., and de Passille, A. M. (2004). Vigilance as a measure of fear in dairy cattle. *Appl. Anim. Behav. Sci.* **87**, 1–13.

Worksheet 8.1 Ethograms.

Behavior	Number of times observed
1	
2	
3	
4	
5	
6	
7	
8	
9	
10	

Detailed descriptions of each behavior:

1.

2.

3.

4.

5.

6.

7.

8.

9.

10.

Worksheet 8.2 Time budgets.

Animal 1

Behavior	Cumulative Time
1	
2	
3	
4	
5	
6	
7	
8	
9	
10	

Animal 2

Behavior	Cumulative Time
1	
2	
3	
4	
5	
6	
7	
8	
9	
10	

Animal 3

Behavior	Cumulative Time
1	
2	
3	
4	
5	
6	
7	
8	
9	
10	

Animal 4

Behavior	Cumulative Time
1	
2	
3	
4	
5	
6	
7	
8	
9	
10	

Animal 5

Behavior	Cumulative Time
1	
2	
3	
4	
5	
6	
7	
8	
9	
10	

Animal 6

Behavior	Cumulative Time
1	
2	
3	
4	
5	
6	
7	
8	
9	
10	

Worksheet 8.3 Grid for recording spatial distribution of vigilant and non-vigilant animals.

You can place flags in your animals' area corresponding to grid points, or simply use approximations based on the animals' relative locations. Mark a "V" for each observation of vigilance, and NV for each non-vigilant animal.

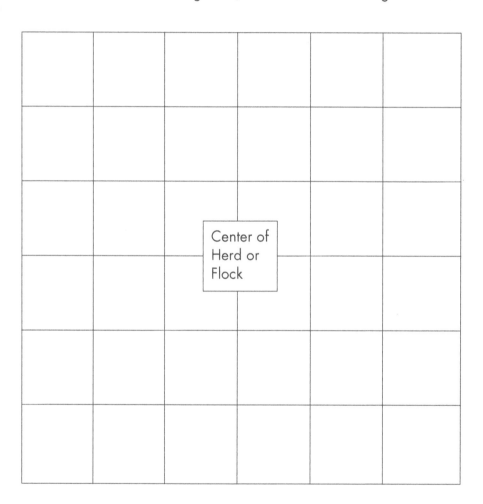

Center of Herd or Flock

Section 3
Laboratory Studies in
Animal Behavior

Chapter 9
Game theory

When faced with several possible ways of behaving, why does an animal behave as it does? One approach to answering this central question is the use of game theory (Axelrod, 1984; Dugatkin and Reeve, 2000). Animal behaviorists construct models that predict how animals might behave as if the animals were playing a game with rules and an objective. These games provide a theoretical context for us to look at animal behavior and to test the robustness of these models. While it is valuable to experimentally corroborate game theory models, it is perhaps even more exciting to find cases where game theory models fail to accurately predict animal behavior. In such a case, we have discovered a system that demands further study to understand when and why these animals are behaving in an apparently aberrant manner.

In this lab we'll explore an interesting game called the Prisoner's Dilemma. The principle of the game is deceptively simple – it simply gives you the options of cooperating with another animal and receiving a small payoff or defecting and potentially receiving a large payoff. Defection, though, carries the risk of punishment, as well as reward. The game is called the Prisoner's Dilemma because it models how prisoners in adjacent cells might behave.

The premise: two miscreants are captured and accused of armed robbery. The suspects are found in possession of illegal weapons, but do not have the cash; they've hidden it. The suspects are placed in separate interrogation rooms. The DA makes the following proposal to both prisoners: "If you confess and your partner does not, you will go free while she goes to prison for 10 years for armed robbery; likewise, if your partner confesses and you do not, you will go to prison for 10 years while she goes free. If both of you confess, I will ask for a reduced sentence of 6 years. If neither of you confess, I only have enough evidence to convict you of illegal arms possession, with a sentence of 2 years." What should the prisoners do? Should the prisoners defect against their partner in crime, or should they cooperate by not confessing? What would you do?

The example of two prisoners deciding whether to squeal on their partner is an example of the game played in a single round — each player only has one decision to make. However, the prisoner's dilemma can also be played a predetermined number of rounds, or it can be iterated — that is, the game has multiple rounds, but neither player knows what round will be the last. As you will see, these distinctions have important consequences for how players should act.

Single play: Consider the payoff matrix in Figure 9.1. If you were player A, you could get either 3 or 0 points by cooperating with your partner (e.g. not confessing), an average of 1.5 points. You could get 1 or 5 points by defecting (such as squealing), an average of 3 points. Thus, you do better on average to defect if you know you will only interact with this opponent once. Presumably your opponent knows this as well, making it an even worse decision to cooperate in the single play version of the Prisoner's Dilemma. Therefore, the solution to this form of the game is always defection.

What happens if you know that you will interact five times with your opponent? Let's skip to the last move; you are essentially in the same situation as the single play game because there is no future possibility of cooperation with this individual. Both of you know this, and are forced to defect. Thus, the last move where you actually have a decision to make becomes move #4. But, you know that move #5 is already essentially decided, and there will be no cooperation, so again, it is best to defect on move #4. The same logic holds for moves 3, 2, and 1. Thus, when both players know the number of interactions, the solution is the same as in single play — always defect.

What if you don't know how many times you will interact with your opponent? Clearly, both would do better by always cooperating than by always defecting, or even by alternating defection and cooperation. If there is a chance of future interaction, then it may pay to attempt to cooperate with your opponent.

In each turn a player can either cooperate or defect. Players can change their strategies, depending on the previous actions of the other player. At the end of the game, success is measured in the accumulated payoffs of the players. The player with the most points wins. For the purposes of a biology lab, you might want to think of points equaling offspring. Thus, players with more points are reproductively more fit.

You might think that the game is easy, with a single solution, such as predictable strategy of alternating cooperation and defection. As you play the game, though, you'll learn that you need to be clever (or devious) in order to best your opponent. Using contingent strategies, in which you change your behavior depending on how your opponent has played, may become important.

Of course, not all animals bring the same sensory, calculation, memory, and cognitive skills to the table as do humans. Some animals may actually do better than humans, but other animals may remember less, have less insight into their opponent's thinking, or otherwise have limitations that affect their decision-making. When you're done with this lab, don't assume that all animals can play this game, or other behavioral games, in the same way we play it; rather, recognize that the predictions of game theory are hypotheses about how animals might behave. If animals don't behave in accordance with theory, this simply suggests that the theory makes false assumptions about the animals' abilities; for the experimentalist, discord between theory and reality is an exciting invitation to discover how animals really work.

To a certain extent, evolution can compensate for limitations by dictating "programs" of behavior, such as the use of specific strategies in encounters with other animals. Programmed behavior serves many animals well, particularly if they live in predictable environments, in which the same strategy works, generation after generation. Strictly programmed strategies don't fare as well as flexible strategies in a more changeable environment.

I Goals

- To gain an understanding of how learning and/or evolution may shape an animal's behavioral strategies
- To learn how to use theoretical approaches, such as game theory, to form hypotheses and predictions about behavior

II Methods: playing the Prisoner's Dilemma

The Prisoner's Dilemma is a two-player game in which the object is to gain more points by inducing your opponent to cooperate with you.

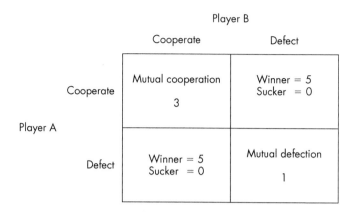

Figure 9.1 Sample payoffs in a Prisoner's Dilemma game.

The table in Figure 9.1 shows the payoffs for the two players, given the choice of cooperating or defecting for each player. If both cooperate, each gets 3 points. If both defect, each gets 1 point. The highest payoff, 5 points, comes if your opponent cooperates and you defect. In this case your opponent gets the "sucker's payoff" of 0. The players do not know the number of turns in the game, although you do know that at some point the game will be stopped.

Think about playing the game. Can you devise a strategy (a combination of cooperation and defection) that will induce your opponent to cooperate?

After you've thought about it a while, check these strategies.

Some examples of Prisoner's Dilemma strategies:

1. RANDOM – This strategy is unpredictable; you randomly cooperate or defect. Your opponent cannot predict your behavior.
2. ALWAYS DEFECT – Your action here is obvious.
3. ALWAYS COOPERATE – Your action here is also obvious.
4. TIT FOR TAT – Start with cooperation on the first move. From then on, always do exactly what your opponent did on his/her previous move.
5. LAST MOVE – What is your best strategy if you know the next turn is the last turn of the game?

Many other strategies are possible; can you figure out other strategies that might give you a competitive edge over the ones we've listed here?

During lab you'll play the game against a few different opponents. Explore the strategies (and come up with strategies of your own) by trying them against different opponents. What strategy works best for you in terms of gaining the highest score? Does the success of your strategy depend on who your opponent is, or does one strategy work equally well against all opponents? Your instructor may organize your class to play a Prisoner's Dilemma tournament; in the tournament, students play against each other in pairs. The winner of each pair goes on the next round, until there is only one student left. What strategies did the winner of the tournament employ?

III Interpretation

Answer the questions on Worksheet 9.1. Were certain strategies consistently successful, regardless of the strategy of the opponent? Did non-verbal signals (your opponent's body language and facial expression) affect how you played the game? What might happen if you played the game through intermediaries, so that non-verbal signals didn't play a role; do you think the lack of these cues would affect your strategy?

References and suggested reading

Axelrod, R. M. (1984). "The Evolution of Cooperation." Basic Books, New York.

Bisazza, A., De Santi, A., and Vallortigara, G. (1999). Laterality and cooperation: mosquito fish move closer to a predator when the companion is on their left side. *Anim. Behav.* **57**, 1145–1149.

Davis, M. D. (1997). "Game Theory, A Nontechnical Introduction." Dover Publications Inc., Mineola, New York.

Dugatkin, L. A. (1997). The evolution of cooperation. *Bioscience* **47**, 355–362.

Dugatkin, L. A. and Wilson, D. S. (2000). Assortative interactions and the evolution of cooperation during predator inspection in guppies (*Poecilia reticulata*). *Evol. Ecol. Res.* **2**, 761–767.

Dugatkin, L. A. and Reeve, H. K. (Eds.) (2000). "Game Theory and Animal Behavior." Oxford University Press, New York.

Also, these web sites give more detail on game theory and the Prisoner's Dilemma:

http://william-king.www.drexel.edu/top/eco/game/game.html

http://plato.stanford.edu/entries/prisoner-dilemma/#Evolution

http://www.princeton.edu/~mdaniels/PD/PD.html

For another interesting game involving human altruism, see:

http://asab.icapb.ed.ac.uk/exercises/alevel_psych/human_altruism.html

Worksheet 9.1 Prisoner's Dilemma.

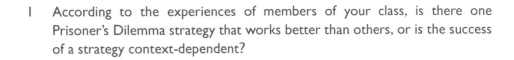

1   According to the experiences of members of your class, is there one Prisoner's Dilemma strategy that works better than others, or is the success of a strategy context-dependent?

2   The Prisoner's Dilemma is presented as an analogy for the evolution of animal behavior. Is it an accurate analogy? Why or why not?

Is it a useful analogy? Why or why not?

In what contexts do you think animals, other than humans, might employ strategies derived from the Prisoner's Dilemma game?

3   Consider a population of animals all of whom employ the ALL D strategy; in other words, they never cooperate. Imagine then a handful of individuals in this population who adopt a TIT-FOR-TAT strategy. Is it possible for TIT-FOR-TAT to ever take over this population? Develop a mathematical model to describe this process.

In this model, the "scores" attributed to each outcome are analogous to what element of evolution?

4   Can ALL D successfully invade a population of all TFT? Why/why not?

Chapter 10
Habitat preferences and choice behavior

How does an animal find the right environment, or niche? Niche is a key concept in ecology (Leibold, 1995; Odling-Smee et al., 2003). Animals express their requirements for existence by seeking out appropriate locations in the environment; for animals, behavior plays the major role in finding their habitat.

In this laboratory we explore how search behavior and microclimate preferences come together so that simple animals can find their way in complex environments. We'll explore the idea that an animal should carry with it "decision rules" about how to behave as it moves into and out of patches of the right microclimate (this could also apply to patches of food). While the rules could be either innate (genetically coded) or learned (the result of experience), the end result is to keep the animal in a location where it has the best chance of survival.

In addition to exploring how animals end up in appropriate habitats, we use this lab to encourage you to develop your abilities make observations, to turn scientific questions into hypotheses, and to design experiments that test your hypotheses. At the outset we give you some background on orientation mechanisms used by animals. We then introduce the specific goals for this lab, a plan of attack that culminates in an oral presentation of your results, and some possible hypotheses that you might test.

The easiest way for an animal to arrive in an appropriate habitat is for it to start moving when conditions are less than optimal and to stop moving when conditions are more appropriate. But should the animal move in a straight line, or should its movements be circular or looping? Thinking this through, you'll realize that movement in a straight line is more likely to take an animal to different environmental conditions. Looping movements are more effective when an animal is searching for something that is nearby. If an animal that needs to be in a shaded area to survive drifts or is blown into a sunny area, it may find the shady area again by looping around. If an animal has found a food item and either experience or evolution tells it that food items occur in groups (patches) then looping movements are an effective way to find the next piece of food. Thus we'd predict that the simplest solution for habitat choice would be for animals to move in a straight line when they're in an inappropriate habitat, to stop when they're in an appropriate location, and to loop if appropriate conditions are likely to be nearby. All of this is done without being able to sense the direction from where the animal is to its goal. The movements that result from this rule, which calls for an animal to stop or start movement depending on conditions, are called *kineses*.

We know from our own experiences that it is sometimes possible to identify the source or direction of a stimulus, and to move relative to the source or direction. As a simple example, place a source of light, such as a flashlight, near pill bugs, sow bugs, or mealworms (you'll have these on hand for this lab). Does the animal move toward or away from the light? Generally pill bugs, sow bugs, or mealworms prefer dark places, so you'd expect directed movement away from the light. This is a *taxis*, because the movement is directed relative to the stimulus. There is a lot of jargon that describes taxes. The pill bug is *negatively phototactic* – moving away from light. A moth that flies upwind is *positively anemotactic*. The beetle that climbs up a grass stem is *negatively geotactic*. And so on.

Humans can use sight and sound to locate stimuli, so our responses to visual and auditory stimuli are usually taxes. This is because our paired sense organs for sight and smell are far enough apart that we can triangulate on the source of a stimulus. However, our sense of smell relies basically on one area of receptors (inside the nose), so you can't locate the source of an odor without moving your head around, and in many cases even then you won't know from where the odor is coming; human responses to odors are often kineses.

| Goals

- To observe taxes and kineses and their associated behaviors in a laboratory setting
- To determine habitat preferences of sow bugs, pill bugs, or mealworms
- To gain experience in developing hypotheses, designing and conducting experiments, and presenting results

II Plan of attack

1. Choose, with lab partner(s), an experimental question from the list below.
2. Based on your choice of question, develop experimental hypotheses and then design experiments to test those hypotheses. Experimental designs should include appropriate controls and a plan for graphing data. Your instructor can help you with statistical tests.
3. Develop a structured plan for recording your data; the worksheets used in other labs in this manual provide good models for doing this. Formatting your datasheets as tables is an excellent approach.
4. Once you've collected the data, the group should graph and analyze it. Prepare a short (five slide) PowerPoint presentation based on your findings. In order, the slides should be: Title, Question and hypotheses, Methods, Graph of results, and Conclusion. At the end of the lab period, your group will present your findings to the class.

III Questions and hypotheses

For this lab, you are asked to form your own hypotheses, but you can work from the following list of questions. After these questions there are sections of this manual chapter that give some natural history background on the test species and some suggestions about experimental design.

1. What are the preferences of terrestrial isopods or mealworms for humidity (damp versus dry), light (dark versus light), and shelter (in the open, or under something)?
   - Are there differences in preferences between roly-polies (*Armadillium*) and sowbugs (*Porcellio*)?
   - Given pair-wise choices between combinations of habitat characteristics, such as sheltered-dry versus sheltered-damp, unsheltered-dry versus unsheltered-damp, can you develop a ranking of which condition – humidity, shelter, or light – has the highest priority for these animals?
   - Based on their pair-wise preferences, what would you predict the most preferred habitat combining all three characteristics would be? The least preferred? Can you test your prediction?
   - Will pill bugs, sow bugs, or mealworms leave a suboptimal environment and pass through an even less optimal environment to search for a more suitable habitat?

2. Based on quantification of pill bug, sow bug, or mealworm movement paths:
   - Can you hypothesize what their decision rules are as they are searching for an appropriate habitat?
   - Do pill bugs, sow bugs, or mealworms use kineses or taxes to orient to appropriate habitat conditions?
3. Are pill bugs, sow bugs, or mealworms positively or negatively phototactic? Are they positively or negatively geotactic?
   - In open field conditions?
   - In T-mazes?
   - How do geotaxis and photaxis interact? In other words, if the animal moves away from light and towards gravity, will it still move towards gravity if the light is coming from the same direction as the force of gravity?
4. If animals in a species are truly cryptic, you would expect them to settle or rest on a background color that matches their body color.
   - Can they do this? What preference do they express when given a choice between a background that matches their body color (tan or brown) and a contrasting color, such as red or green?
   - What does your result suggest about the nature of color vision in this species?
   - Do aposematically colored animals, such as milkweed bugs or hornworm caterpillars, behave as predicted and seek contrasting backgrounds?

IV The test organisms, questions, and hypotheses – background information

A Habitat preferences in terrestrial Crustacea or mealworm larvae

Land-living Crustacea, known as roly-polies, pill bugs, sow bugs, and woodlice (http://bugguide.net/node/view/15976/bgpage) are common through most of North America and serve as a excellent model for studying habitat choice. Like their aquatic relatives, terrestrial Crustacea have gills that must be kept moist for them to be able to exchange oxygen and carbon dioxide with the atmosphere. These terrestrial Crustacea are in the class Isopoda, so named because their legs are all roughly the same size and shape.

Pill bugs (*Armadillium vulgare*) are black and are capable of rolling into a tight ball when threatened (hence, their other common name, rolypolies). Sow bugs (genera *Oniscus* and *Porcellio*) are gray and don't display the same interesting defensive behavior. Females of both pill bugs and sow bugs carry their eggs and, after hatching, young, in a pouch called a "marsupium." *Armadillium vulgare* and *Porcellio* species are both introduced into North America, presumably from Europe, and have become widespread and common in their new habitats.

Pill bugs and sow bugs forage on decaying plant material and prefer to live in moist, protected, dark locations; they are typically found under rocks and logs. They may leave protected areas under the cover of darkness to feed. In the lab, pill bugs or sow bugs can be maintained on a layer of moist dirt or sand, under damp organic material, such as leaves. If fed small pieces of vegetables, such as carrot or potato, they will reproduce and thrive in captivity (http://www.colostate.edu/programs/pestalert 17:10 page 3, 2000).

Mealworms are larvae of a grain beetle, *Tenebrio molitor*. The larvae eat milled grain and have been used as model organisms in laboratory studies of ecology (Howard, 1955).

These natural history observations lead to simple hypotheses concerning pill bugs', sow bugs', or mealworms' movements. Under unfavorable conditions (dry, high light intensity) we would expect the crustaceans to continue moving, while under favorable (moist, dark) conditions we would expect movement to stop. This is a simple enough rule, but is it adequate to explain how pill bugs optimize their habitat choice?

Some questions can be answered by putting damp filter paper on one half of a Petri dish and dry filter paper on the other half. Do the crustaceans prefer the dry half or the wet half? How long does it take them to settle in a half? Are they more likely to move if they start in the wet half or in the dry half? Alternatively, you can put an animal at the midpoint of a 10-cm-long tube, plug one end with wet cotton and the other with dry, and observe its preference for location in the tube. Your data could consist of either counts (one count per replicate) of which side of the Petri dish or which end of the tube the animal is at after a fixed period of time, such as five minutes. Alternatively, you could measure the percentage of time each animal is on each end or side of the experimental apparatus.

An important feature of experimental design is that it is not a good idea to put several animals in the test container at the same time. If you do this, the behavior of one animal can influence the behavior of others;

Table 10.1 Relative humidity above saturated solutions of selected salts (Winston and Bates, 1960).

Compound or salt-relative humidity above saturated solution (%)	
Sodium hydroxide (NaOH)	5
Glucose	55
Sodium chloride (NaCl) plus sucrose	63
Sodium chloride (NaCl)	75
Sucrose	85
Potassium sulfate (K_2SO_4)	97

Caution should be used with the sodium hydroxide solution, which is quite caustic. The other solutions are relatively benign. Humidities given in the table are the approximate result at room temperature.

the result is that the animals do not behave independently of each other. If you put five animals in the test apparatus, you cannot count each animal separately when you score your data. It is simpler and experimentally more appropriate to test only one animal at a time.

You may have a more sophisticated way of testing humidity preferences available to you. This is an apparatus that allows pill bugs, sow bugs, or mealworms to settle in an area of preferred humidity. The pill bugs, sow bugs, or mealworms can move freely on walkways that are held above salt solutions. Winston and Bates (1960) discovered that the humidity of the atmosphere above a saturated salt solution depends on the kind of salt used to make the solution. This discovery has given ecologists and behaviorists a simple tool for giving animals habitat choices based on humidity. We've listed a few saturated solutions (Table 10.1) to produce a range of humidity conditions. Given the choice between the airspace over two of the solutions, which one do the crustaceans choose? Once you know the preferred relative humidity, you can then ask whether the crustaceans will move through an unfavorable condition to get from a suboptimal to an optimal humidity.

B Phototaxis and geotaxis in terrestrial Crustacea or mealworm larvae

There are two useful experimental designs for exploring taxis in response to stimuli such as light or gravity. The first is an open field design, in which the animal is allowed to move freely in an arena and then the stimulus is applied directionally to the animal, so that its response can be measured. For example, a beam of light could be directed at a right angle to the animal's direction of travel. Does it turn toward or away from the light source? The answer to this question establishes whether it is phototactic, and if so, if the phototaxis is positive or negative. A similar choice relative to gravity could be achieved by tipping the arena so that the animal has the opportunity to go up or down.

The second sort of design uses a simple T-maze. Constructed using plastic or glass tubing, the animal walks through a section of tubing until it reaches a branch point or T. It then must choose which direction to turn. If a stimulus, such as light or an odor, is applied from one direction, then the animal's tendency to move toward, or away from, the stimulus can be scored. Open field arenas allow for more realistic movements of the animal, but a T-maze makes it possible to control animals that might fly away, and it makes sensory inputs, such as odors, easier to manage.

Both open field and T-maze approaches can be used in assessing responses of pill bugs, sow bugs, or mealworms to light and gravity. How do these approaches compare? What can you conclude about the responses of this species?

C Are cryptic animals able to match their color to a background?

Natural history accounts often focus on marvelous examples of cryptic coloration. Pit vipers, venomous snakes such as rattlesnakes and copperheads, are often near-perfect matches for the dried leaves or soil; their cryptic coloration aids their sit-and-wait hunting strategy. You've probably seen moths that are perfect matches for bark or dead leaves. Another oft-cited example, chameleons – color-changing lizards – probably change color not to match their background, but to absorb or reflect heat, or sometimes the changing colors are social signals to other lizards.

But coloration is not always cryptic. In some cases animals – generally poisonous or stinging species – advertise their presence with bright or contrasting colors. Coral snakes are a great example of this, as are yellow jackets, hornets, and honeybees. This is termed aposematic coloration, and to stand out, we'd predict that aposematically colored animals will settle on a background that maximizes their visibility.

For this part of the lab, the organisms available will depend on the season. Among the organisms that could be tested are crickets, grasshoppers, milkweed bugs, mantids, crab spiders, and moths.

For a simple start on this question, cover half of the bottom of a container with black construction paper and the other half with white. Place an animal in the container and observe whether it prefers the dark or white background. Observe and record the strategies used by the animals to find their preferred background. Count the number of trials in which each animal is on each background at the end of a fixed period of time, or the percentage of time each animal spends on each background.

If adult grasshoppers are present outdoors when you are doing this lab, you can extend this study by walking through a field and observing where grasshoppers land when they jump after being disturbed. Do they land on matching or contrasting backgrounds? Can you devise a way of quantifying this result?

References and suggested reading

Anon. (No date). http://entomology.unl.edu/k12/isopod.shtml

Anon. (2003–2007). http://bugguide.net/node/view/15976/bgpage

Berg, C. J. (1975). Orientation to physical conditions by terrestrial isopods. In "Animal Behavior in the Laboratory and Field" (Price, E. O. and Stokes, A. W., Eds.). W.H. Freeman and Company, San Francisco, pp. 46–48.

Brown, C. (2006). Taxis in animals. http://iweb.tntech.edu/cabrown/AnimBehav/AB%20Labs.htm

Cuadrado, M., Martin, J., and Lopez, P. (2001). Camouflage and escape decisions in the common chameleon *Chamaeleo chamaeleon*. *Biol. J. Linn. Soc.* **72**, 547–554.

Eterovick, P. C., Figueira, J. E. C., and Vasconcellos-Neto, J. (1997). Cryptic coloration and choice of escape microhabitats by grasshoppers (Orthoptera: Acrididae). *Biol. J. Linn. Soc.* **61**, 485–499.

Hare, J. F. (2005). Animal Behaviour Lab 1: Humidity Preferences of the sowbug (*Oniscus asellus*) http://umanitoba.ca/science/zoology/ faculty/hare/z310/Sowbug.doc

Howard, R. S. (1955). The biology of the grain beetle *Tenebrio molitor* with particular reference to its behavior. *Ecology* **36**, 262–269.

Leibold, M. A. (1995). The niche concept revisited – mechanistic models and community context. *Ecology* **76**, 1371–1382.

Odling-Smee, F. J., Laland, K. N., and Feldman M. W. (2003). "Niche Construction : The Neglected Process in Evolution." Princeton University Press, Princeton, NJ.

Todd, P. A., Briers, R. A., Ladle, R. J., and Middleton, F. (2006). Phenotype-environment matching in the shore crab (*Carcinus maenas*). *Marine Biol.* **148**, 1357–1367.

Winston, P.W. and Bates, D.H. (1960). Saturated solutions for the control of humidity in biological research. *Ecology* **41**, 232–237

Chapter 11
Pheromones

Animals use pheromones, or odor communication, in many different ways (McClintock, 1998; Wyatt, 2003). These include mate attraction, territorial marking, alarm, and communication about the location of food. While pheromones were first discovered in insects, we now know that odors play an important communicatory role in many different types of animals, including vertebrates such as fish, amphibia, reptiles, and mammals. Birds are probably the only major group of animals in which odor communication is, at present, unknown.

Pheromones are produced by exocrine glands – groups of cells that are specialized for odor production. In many cases, the exocrine gland is a modification of dermal, or skin, cells and lies on or close to the animal's surface. Almost any type of organic chemical compound can serve as a pheromone – the only requirement is that the animal must be able to produce the compound and that it must also be perceived by members of that species. Oftentimes, animals produce pheromones in minute quantities, making determination of their chemical identity difficult.

In this lab we'll explore in depth two very different types of odor communication, termite trail pheromones and human gender identification and mate attraction pheromones. We'll also do a brief demonstration of honeybee alarm pheromone. Very different experimental methods are used for these three kinds of pheromones, so that by studying all three you'll get broad experience in how animal behaviorists look at hormones. These three kinds of pheromones also provide excellent examples of the different ways in which animals use odors for communication.

1 Termites

We think of most termites as simple, wood-feeding insects. Living their lives underground in chambers and tunnels in the soil or aboveground in tunnel systems constructed of mud, sawdust, and saliva, many termite species lack eyes and all termites rely heavily on their social group, or colony, for survival. Worker termites are sterile, meaning that they must reproduce indirectly, by aiding the queen and king (who are close relatives to the workers) in rearing their young. Termites are also soft-bodied, lacking a protective covering from heat and sunlight. A few seconds exposure to strong sunlight or high heat can easily kill a termite, a fact that goes along with their preference for dark, confined spaces.

Within the confines of the nest and tunnel system, how do termites communicate? Well, of course this lab is about pheromones, so it doesn't take a rocket scientist to guess that pheromones are a key mode of termite communication, and that a termite pheromone is the subject of this part of the lab.

The key navigational problem for termites is to stay on the correct path leading between the colony and food. Even though the tunnel system provides some guidance, tunnels branch, merge, and ultimately break down into galleries within the wood that the termites are eating. Worker termites must be able to find the foraging route, even if they can't see other termites or landmarks. Odors are the perfect solution.

Pheromone trails, like those used by termites, have one distinct disadvantage. Most odor trails are not polarized, meaning that the odor does not tell the animal anything about which direction on the trail leads to home. This lack of polarization of the trail forces termites to use other cues to determine which way to go, once they've found the trail (Runcie, 1987; Cornelius and Bland, 2001; Reinhard and Kaib, 2001).

We'll use workers of the termite genus *Reticulitermes* to explore how termites use trail pheromones. *Reticulitermes* live underground and are found in all but the most arid or highest elevation regions of the United States (Thorne et al., 1999). This genus is responsible for considerable property damage, as the workers will eat wood used in home construction, and most modern construction includes concrete, metal, or plastic barriers between the wood and the ground to prevent termite infestation.

Reticulitermes workers can be divided into two readily observable classes. Foragers have small, soft heads, whereas soldiers have large, more sclerotized (hardened, darkened exoskeleton) heads and strong mandibles. The sample of workers we'll have in lab will probably be mostly foragers, but may also include some soldiers. In a full colony you would find thousands of workers and a queen and king. The queen is much larger than the workers, and once had wings that she used in her mating flight. After founding her colony, she breaks her wings off, as does her mate. Wings are unwieldy and superfluous in a subterranean environment.

The primary chemical compound in *Reticulitermes* trail pheromone is (Z,Z,E)-3,6,8-dodecatrien-1-ol. Serendipitously, this compound also is present in some ballpoint pen inks, a fact that we'll take advantage of in lab. If you find the right ballpoint pen, you can draw a line and the termites will follow. We'll also have some synthetic trail pheromone compound for you to use. This is a little harder to work with, as you can't see where you've laid it down, but the synthetic compound can be diluted, opening the opportunity to test for how termites respond to trails of different strengths.

What other species use trail pheromones? The most common example is ants; many different species of ants use trail pheromones in the same way that termites do. Stingless bees living in dense tropical rain forests mark leaves with odors, establishing a trail through the trees. In mammals, trail pheromones are unknown but many mammals use scents to mark territory boundaries. The hypothesis that scent marks and/or trail pheromones are used for communication is reasonable in any animal that operates in the dark, or that communicates with animals who may visit the same spot, but at different times.

A Goals

- To observe the importance of pheromones in animal navigation
- To learn the basics of termite biology
- To determine what challenges face animals that use trail pheromones for navigation

B Questions and hypotheses

For this part of the lab, we'd like you to develop your own questions and hypotheses. You might want to think about these questions: Do all termites respond in the same way to odor trails? Do they all have the same strategy for finding a trail when they've lost it? How do termites resolve complications like losing their trail, branches, or forks in the trail, and dilution of the trail due to evaporation? Can directional information be incorporated into an odor trail? Working with a partner, write down two or three hypotheses that you intend to test. Your instructor will collect copies of everyone's hypotheses so that they can facilitate discussion of your results.

C Methods

You will have worker termites, soldier termites, paper, ballpoint pens, and possibly (Z,Z,E)-3,6,8-dodecatrien-1-ol available. If you have the chemical, your instructor will show you how to make a serial dilution of the (Z,Z,E)-3,6,8-dodecatrien-1-ol in a neutral carrier (paraffin oil). The basic strategy for your experiments is to create trails on the paper and then to observe the trail-following response of the termites.

D Interpretation

The instructor will lead you through a discussion of the hypotheses tested by your lab section. Once you've heard about everyone's data, how efficient do you think trail pheromone communication is? What are the advantages and limitations of using odor trails? If you had the role of a "guiding hand" on evolution, how would you modify the termite's use of trail pheromones?

II Honeybee alarm pheromone

Honeybees use a highly volatile chemical, isopentyl acetate (IPA), as the principle factor in their alarm pheromone (Collins and Blum, 1983; Moritz et al., 1985; Lensky et al., 1995; Wager and Breed, 2000). Your instructor will have a small quantity of IPA in a bottle for you to sniff. To most people, IPA has a banana-like scent. Do you recognize it from times that you've been stung by honeybees? The IPA is released, along with other alarm pheromone components, when the bee's stinger is left in your skin. We'll have a small group (5–10) honeybees in a sealed container in the lab. How active are they? What is their response when the instructor puts a small quantity of IPA into their container?

III Human pheromones

A Introduction and background

Humans in many contemporary cultures do their best to suppress their natural odors, so it may come as a surprise to you that communication using odors has, historically, played an important role in the behavior of humans. Studies show that humans can identify their mates and close relatives by using odors, and that odors may have particular importance for newborn infants (McClintock, 1998; Weller, 1998).

In adult humans, odor production is concentrated in obvious places – the axillae (armpits) and the pubic area. The same hormonal processes that govern other aspects of puberty initiate odor production, and in our culture we start to use deodorants and antiperspirants during puberty in order to suppress the odors. Hair growth in the axillary and pubic areas probably serves at least partly to provide increased surface area for odor evaporation.

Women who live together, such as in a dormitory or sorority, often observe that their menstrual cycles become synchronized (Stern and McClintock, 1998; Whitten, 1999). Pheromones play a key role in causing this synchrony. Some women (generally those regarded as socially dominant by their peers) are effective at subconsciously synchronizing the menstrual cycles of other women; the odors that organize menstrual synchrony come from the axillae.

Studies of human pheromones are, because of ethical concerns, non–invasive. The most common technique uses t-shirts, which are worn by the subjects of the experiments and which, by contact with the subject, absorb odors. T-shirt studies have been used to test the hypothesis that newborn infants recognize their mothers by odor (they do!), that spouses can recognize their partners by odor (yes to this also), that facial attractiveness and odor attractiveness are correlated (they are!), and that scents are responsible for menstrual synchronization.

We'll use the t-shirt technique to test our own hypotheses concerning gender identification and attractiveness of odors.

B Goals

- To learn about the importance of pheromones in animal recognition and attraction
- To determine whether odors play a role (perhaps subconsciously) in human communication

C Questions and hypotheses

When would you expect humans to use pheromones? We'll test three hypotheses about human chemical signals:

- First, we hypothesize that you can tell males from females by odor. To further pursue this, we'll use the same data to test for differences in male and female abilities to discriminate gender on the basis of odor.

- Second, we hypothesize that male odors will be more attractive to females and that female odors will be more attractive to males.
- Third, we also hypothesize that female odors will be more attractive to males in the 10 days in the middle of their menstrual cycle (when they are most likely to ovulate) than during the rest of their cycle, and that women using hormonal birth control (or who are pregnant) will have less attractive odors than women who likely are ovulatory.

D Methods: the t-shirt test

A note before beginning. We want to make sure that you do not feel that participating in this experiment compromises your privacy. If you follow the instructions, there will be no way for the t-shirt you use to be matched to you, after you turn it in. If you don't feel comfortable participating, there is no requirement for you to do so, and for women, if you don't feel comfortable answering the questions about your reproductive status, simply leave them blank.

In lab, the week prior to the experiment, you'll be given a clean t-shirt in a ziplock bag. The bag will also contain a card on which you will mark your gender. For female students there will be optional questions about your reproductive status; we want to emphasize that answering these questions is voluntary. Do not put your name on the card; this maintains your anonymity.

Please follow these instructions exactly. On Friday morning you may follow your normal routine of showering or bathing, but try to avoid scented soaps and shampoos, and don't apply perfume, cologne, or other scented products. Do not apply deodorant or antiperspirant. Avoid foods containing garlic or onion. Friday evening take a shower to rinse yourself before going to bed, but do not use soap or shampoo, and do not apply scented products, deodorant, or antiperspirant. You can think of what follows as your stinky weekend.

Put the t-shirt on as an undergarment when you wake on Saturday and wear it continuously until Monday morning. On Saturday morning do not shower or bathe, and do not shower or bathe until you remove the t-shirt on Monday morning. Continue to avoid foods containing garlic or onion during the day and don't apply perfume, cologne, other scented products, deodorant, or antiperspirant on Saturday and Sunday. Try to avoid environments that will affect the smell of your t-shirt, such as smoke-filled rooms. At night, wear the shirt as an undergarment while you sleep. Monday morning, put the shirt back in its ziplock bag with the card, and return to your normal routine for hygiene. Drop the shirt off (leaving it in its plastic bag) in the box in the lab on Monday morning. Remember that all this will have been for nothing if you don't fill out the card and put it in the bag with the shirt, but be sure not to put your name on the card.

The instructor will sort the t-shirts into groups of 10. Some groups will contain both male-exposed and female-exposed shirts, while other groups may contain shirts worn by only one gender. The shirts will be numbered to match the information on the cards and after the instructor records the information, the cards will be removed from the bags and destroyed. During lab you will briefly sniff each shirt in one of the groups and write down the shirt's number, whether you think it was worn by a male or a female, and a qualitative assessment of the attractiveness of the odor, on a scale from 1 (repugnant) to 10 (highly attractive). The score sheet for this is Worksheet 11.1.

After you do the sniff tests, the instructor will have you decode the shirt numbers by giving you a list of which shirts (by number) were worn by males and females. Females who are pregnant or on hormonally based birth control will be marked with an *, and females who are likely to be ovulatory, based on the number days since their last period began, will be marked with a #.

E Interpretation

How well did you do with identifying gender by odor? Were odors of members of the same gender or of the opposite gender (from yours) more attractive? In your lab, your instructor will collate the data and your lab group will discuss the results.

Just in case you didn't believe that you're doing serious science with the t-shirt test, we cite here some recent papers in which scientists used t-shirt tests or similar methods to test hypotheses such as ours about attractiveness and mate choice in humans (Wedekind et al., 1995; Wedekind and Furi, 1997; Gangestad and Thornhill, 1998; Rikowski and Grammer, 1999; Jacob and McClintock, 2000; Morofushi et al., 2000; Shinohara et al., 2001; Singh and Bronstad, 2001; Rantala et al. 2006).

References and suggested reading

Collins, A. M. and Blum, M. S. (1983). Alarm responses caused by newly identified compounds derived from the honeybee sting. *J. Chem. Ecol.* **9**, 57–65.

Cornelius, M. L. and Bland, J. M. (2001). Trail-following behavior of *Coptotermes formosanus* and *Reticulitermes flavipes* (Isoptera : Rhinotermitidae): Is there a species-specific response? *Environ. Entomol.* **30**, 457–465.

Gangestad, S. W. and Thornhill, R. (1998). Menstrual cycle variation in women's preferences for the scent of symmetrical men. *Proc. Roy. Soc. London, Series B, Biol. Sci.* **265**, 927–933.

Jacob, S. and McClintock, M. (2000). Psychological state and mood effects of steroidal chemosignals in women and men. *Horm. Behav.* **37**, 57–78.

Lensky, Y., Cassier, P., and Telzur, D. (1995). The sebaceous membrane of honeybee (*Apis mellifera* L) workers sting apparatus – structure and alarm pheromone distribution. *J. Insect Physiol.* **41**, 589–595.

McClintock, M. K. (1998). Pheromone communication and interaction with hormones – on the nature of mammalian and human pheromones. *Ann. New York Acad. Sci.* **855**, 390–392.

Moritz, R. F. A., Southwick, E. E., and Breh, M. (1985). A metabolic test for the quantitative-analysis of alarm behavior of honeybees (*Apis mellifera* L). *J. Exper. Zool.* **235**, 1–5.

Morofushi, M., Shinohara, K., Funabashi, T., and Kimura, F. (2000). Positive relationship between menstrual synchrony and ability to smell 5 alpha-androst-16-en-3 alpha-ol. *Chem. Senses* **25**, 407–411.

Rantala, M. J., Enksson, C. J. P., Vainikka, A., and Kortet, R. (2006). Male steroid hormones and female preference for male body odor. *Evol. Hum. Behav.* **27**, 259–269.

Reinhard, J. and Kaib, M. (2001). Trail communication during foraging and recruitment in the subterranean termite *Reticulitermes santonensis* De Feytaud (Isoptera, Rhinotermitidae). *J. Insect Behav.* **14**, 157–171.

Rikowski, A. and Grammer, K. (1999). Human body odour, symmetry and attractiveness. *Proc. Roy. Soc. London, Series B, Biol. Sci.* **266**, 869–874.

Runcie, C. D. (1987). Behavioral evidence for multicomponent trail pheromone in the termite, *Reticulitermes flavipes* (Kollar) (Isoptera, Rhinotermitidae). *J. Chem. Ecol.* **13**, 1967–1978.

Shinohara, K., Morofushi, M., Funabashi, T., and Kimura, F. (2001). Axillary pheromones modulate pulsatile LH secretion in humans. *Neuroreports* **12**, 893–895.

Singh, D. and Bronstad, P. M. (2001). Female body odour is a potential cue to ovulation. *Proc. Roy. Soc. London, Series B, Biol. Sci.* **268**, 797–801.

Stern, K. and McClintock, M. K. (1998). Regulation of ovulation by human pheromones. *Nature* **392**, 177–178.

Thorne, B. L., Traniello, J. F. A., Adams E. S., and Bulmer M. (1999). Reproductive dynamics and colony structure of subterranean termites of the genus *Reticulitermes* (Isoptera Rhinotermitidae): a review of the evidence from behavioral, ecological, and genetic studies. *Ethol. Ecol. Evol.* **11**, 149–169.

Wager, B. R. and Breed, M. D. (2000). Does honeybee sting alarm pheromone give orientation information to defensive bees? *Ann. Entomol. Soc. Am.* **93**, 1329–1332.

Wedekind, C and Furi, S. (1997). Body odour preferences in men and women: do they aim for specific MHC combinations or simply heterozygosity? *Proc. Roy. Soc. London, Series B, Biol. Sci.* **264**, 1471–1479.

Wedekind, C., Seebeck, T., Bettens, F., and Paepke, A. J. (1995). MHC-dependent mate preferences in humans. *Proc. Roy. Soc. London, Series B, Biol. Sci.* **260**, 245–249.

Weller, A. (1998). Human pheromones – Communication through body odour. *Nature* **392**,126–127.

Whitten, W. (1999). Reproductive biology – Pheromones and regulation of ovulation. *Nature* **401**, 232.

Wyatt, T. D. (2003). "Pheromones and Animal Behaviour: Communication by Smell and Taste." Cambridge University Press, New York.

Turtlecoaster for Snails!

Worksheet 11.1 Human pheromone lab   (no need to put your name on this form).

Please record the following information:

Your gender:_____

Fill out the first three lines BEFORE looking at the key.

Shirt Number									
Gender (your guess)									
Attractive ness Score (1–10)									
Actual Gender (from key)									

Then calculate for your sniffing results:

Number of correct gender ID's for shirts worn by males	Number of correct gender ID's for shirt worn by females	Number of incorrect gender ID's for shirts worn by males	Number of incorrect gender ID's for shirts worn by females	Average attractiveness score for shirts worn by males	Average attractiveness score for shirts worn by females

Does ovulatory status affect attractiveness of female odors?

Average attractiveness score of * females	Average attractiveness score of of # females

Chapter 12
An experimental exploration of *Betta* behavior

This chapter contains three experiments using Siamese fighting fish, *Betta splendens*. (Note that Chapter 13 also uses *Bettas*.) The following information on Siamese fighting fish, *Betta splendens*, will help in your understanding and interpretation of the experiments on *Bettas*; if you explore all of the questions suggested in this chapter you will spend several weeks with these fish. We hope that you will enjoy the time spent with this beautiful and fascinating species.

We think that pursuing the behavior of one species, in-depth, has distinct advantages. First, of course, you gain deeper familiarity with the biology of this species than you would if we only spent one week with it. Second, by using the same animals in several laboratories we minimize the impact of our animal usage. *Bettas* are bred in captivity, so we are confident that no habitat destruction is associated with their collection. This program of reuse minimizes the total number of animals used during the semester. Third, by observing the behavior of this species in different contexts — learning, territoriality, mate choice — you gain a fuller understanding of how complex and sophisticated even a "lower" vertebrate species can be.

It is always a good idea to understand the basic biology of any animal you may wish to study. This knowledge helps guide the types of questions you may experimentally ask of your animal, as well as the types of methods that are appropriate. Here is the background information on the natural history, behavior, physiology, and genetics of *Betta splendens*, the Siamese fighting fish.

The genus *Betta* consists of over 100 species, all of which are native to the Southeast Asian countries of Thailand, Vietnam, Cambodia, Malaysia, and Indonesia (Goldstein, 2001). For centuries, these fish have been bred for their fighting ability, their tail shape, and their coloration. This artificial selection has resulted in an exaggeration of males' willingness to fight, and has given rise to elaborate tails and dazzling body colors.

Bettas belong to a group of fish known as the Anabantoids, or labyrinth fish. Fish in this group possess a vascularized first gill arch near the mouth. Through evolutionary time, this vascularized region became increasingly convoluted, providing a larger surface area across which gas exchange could occur. This unique organ allows *Bettas* and other labyrinthine fish to gulp and extract oxygen from the air. When one considers that air is 2100 ppm (parts per million) oxygen (O_2), while water is only 8 ppm O_2, the advantage to being able to extract O_2 from the air becomes apparent. The labyrinth organ has allowed *Bettas* to thrive in swampy, stagnant water where other fish could not survive. When you see your *Betta* behave as if it is gulping air, you will know that it is indeed taking a breath!

When another male confronts male *Betta*, they exhibit several character-istic agonistic behaviors. These include flaring of the fins, rapid undula-tion, and expansion of the operculum (gill membrane) and underlying brachiostegal membrane (See Figure 12.1). You will become very famil-iar with these behaviors over the course of the *Betta* experiments. As such, we will not divulge more information here about these behaviors – that's for you to discover!

If you own *Betta splendens*, one day you may see your male blowing small bubbles and creating a mat of these bubbles on the surface of the water. What is he doing? Chances are, your male *Betta* is feeling paternal; he is creating a bubble nest in which the young *Betta* fry will develop. The eggs and young fry do not have developed gills or labyrinth organs; instead, they absorb O_2 directly through their skin. The bubble nest serves to suspend the young near the surface where O_2 concentration is the highest. The bubble nest probably also serves a protective function, both from pathogenic disease and from other predators. The bubbles

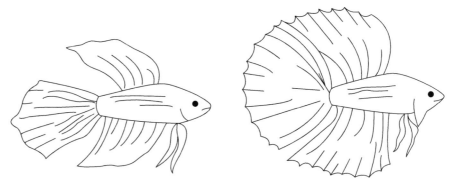

Figure 12.1 Contrast between male *Betta* in normal swimming posture and male in display posture. Note the larger appearance, gained by flaring the fins and the operculum.

are coated with some of the male's own slime coat, which helps ward-off disease (Jaroensutasinee and Jaroensutasinee, 2001).

Female *Bettas* exhibit outward signs of sexual receptivity. When she displays horizontal lines, seems uninterested in the male, and has no white oviduct spot, she is not ready to breed, and the male will often behave aggressively toward her. She displays vertical bars and actively seeks the male when she is ready to spawn. The male herds the female toward his bubble nest, where he curves his body around hers. She releases a stream of eggs that pass through a cloud of his sperms as the eggs sink. The sperms adhere to the surface of the eggs and fertilize them. The male recovers first and catches the eggs as they sink. He gathers them in his mouth and expels them into the bubble nest. The female munches on the remaining eggs, and the couple repeats the entire spawning process up to 50 or 60 times, producing up to 15 viable eggs with each bout. After the female has expelled all of her mature eggs, the male remains excited and is ready for more spawning. At this point, the spent female must make a hasty retreat or risk severe harassment from the randy male.

After the female escapes, the male begins care of the nest. He replaces bubbles as needed, and cleans the eggs by rolling them around in his mouth. After a few days, the eggs will have hatched and developed into motile larvae. The larvae stay near the nest, but at this point the male's attention begins to wane. After another day, the larvae begin to feed, and the male swims away in search of his own meal.

Betta fin morphology and coloration are under genetic control. Fin length is under dominant/recessive gene control – long fins are dominant – while coloration is controlled by incomplete dominance. The incomplete dominance of the color genes helps explain the large variation in *Betta* color, from green, to blue, to red, to black, and combinations of these colors. *Betta* sex determination appears to follow the XX female/XY male pattern, as in humans, with an interesting twist; females who are partially ovariectomized can become sexually functional males!

In the subsequent *Betta* exercises, we will explore aspects of male/male aggression, male/female interactions, and female mate choice. We hope that this section has provided you with enough background information that the results of our exercises will make sense. If you are interested in delving deeper into *Betta* biology, there are a number of good books, articles, *Betta* clubs, and *Betta* websites.

I Territoriality in *Betta splendens*

Males of many species establish defended spaces or territories (Forslund, 2000; Switzer et al., 2001; Peake et al., 2006). These territories typically include some critical resources such as food, water, or shelter. Males are willing to aggressively defend their territories against interlopers, and they maintain their borders with surrounding males' territories. Male defensive displays are typically stereotyped, ritualized behaviors.

Most of the time, aggressive interactions between a territorial male and an intruding male end before they rise to the level of actual physical harm to either animal. Males usually are able to assess the fighting ability of their opponent, and the likelihood of defeating him, without actually resorting to all-out combat. Exaggerated display behaviors by the males often are enough for both to discern the probable winner. In cases where neither male is clearly better, the interaction may progress toward heightened aggression and actual violent combat.

The display behavior of the male *Betta* is an example of a fixed action pattern. The signal – the appearance of an intruder – prompts a series of ritualized behaviors in the subject male. Because all male *Bettas* perform these behaviors, there is no ambiguity as to their meaning. As in male/male interactions in other species, these behaviors allow each male to "size-up" his rival and decide whether he should continue to escalate the display, or to disengage. In today's exercise, we will observe male/male territorial displays in males shown to other males, and shown to their own reflection. Take a moment to consider how you might expect males to behave in both contexts.

A Goals

- To learn to identify the behaviors associated with male/male territorial aggression in *Betta splendens*
- To determine experimentally when and why males escalate their aggressive displays

B Questions and hypotheses

Consider the information about conflict escalation in the introduction to formulate a hypothesis about how male *Bettas* will react to other males, and how they will react to their own reflection.

C Methods

Initial observations

Place two male *Bettas'* tanks next to each other for one minute and observe the variety of agonistic behaviors each male displays. Record the qualitative observations of the interaction. After the one-minute trial, decide how you will quantitatively record data for all subsequent trials. Create a data sheet appropriate for the experimental manipulations and the types of data you will collect.

The experiment

All males have been kept visually isolated for over one week. We will perform brief introductions of "intruder" males, but it is critical to keep the males visually isolated during times that trials are not in progress.

1. Observe the number of aggressive behaviors your male displays in one minute while he is visually isolated.
2. Place the tanks of two males next to each other and remove the visual barrier between them. Record the aggressive behavior of both males for one minute. Replace the visual barrier and allow both males to remain visually isolated for at least one minute before using them in subsequent trials.
3. Place a mirror next to the male's tank and record the aggressive behavior for one minute. Allow the male to rest for one minute after this trial.
4. Repeat 1–3 for new male *Bettas*.

D Interpretation

Was your hypothesis supported by the results of the experiment (Worksheet Figure 12.3)? What does the result from this experiment tell us about aggression in animals, in general? Which treatment had the highest level of aggression, and why do you suppose this treatment had the highest level of aggression? What is the purpose of observing aggressive behavior of males that are visually isolated? How do you suppose artificial selection might have altered the outcome of this experiment? Later in the semester, we'll learn that some investigators argue that animals that can identify "self" in a mirror have cognitive abilities. Did you see any evidence of recognition of self in *Bettas*?

II Testing the dear enemy hypothesis in *Bettas*

In the previous exercise, we observed male *Betta* territorial behavior. It should be clear that the behaviors displayed by the males are energetically costly. This raises the question of how these males might behave in nature when they encounter other males. When male *Bettas* establish territories, they are likely to have neighbors with territories adjacent to their own. This could set up a situation in which males are constantly defending their territory against other males. However, this is not the case. Males will often tolerate the presence of neighbors. Why is this? In today's exercise, we will test two proposed hypotheses for why male *Bettas* tolerate the presence of neighboring males (Hojesjo et al., 1998; Leiser and Izkowitz, 1999; Utne-Palm and Hart, 2000).

The first hypothesis is the "habituation hypothesis", which states that males who possess a territory surrounded by other males' territories get accustomed to the presence of other males and become desensitized to their presence. The second hypothesis is the "dear enemy hypothesis"; males that possess a territory surrounded by other males' territories learn the identities of their neighbors. These familiar males are not a threat because they are already territory holders. You will notice that the key difference between the hypotheses lies in whether or not *Bettas* have the capacity for individual recognition.

A Goals

- Be able to identify the behaviors associated with male/male territorial aggression in *Betta splendens*
- Experimentally test the habituation hypothesis versus the dear enemy hypothesis in *Bettas*

B Questions and hypotheses

The two main hypotheses for this exercise were described in the introduction. Make predictions of how you would expect the males to behave in each different pairing that support (a) the habituation hypothesis (and b) the dear enemy hypothesis.

C Methods

We will have two groups of male fish. In one group, the males have been kept in tanks visually isolated from all other male *Bettas* – these are the isolated males. In the other group, males have been kept in tanks adjacent to two other male *Bettas'* tanks – these are the community males.

There are three important pairings to perform. The first is to present the isolated male with a novel intruder male fish. The second is to present a community male with a novel intruder male fish. The third is to separate for a minute two community males who have been stored together, and then place these two males back together. Record the agonistic behavior for each pairing in Table Worksheet 12.1. Repeat each pairing with new subject males.

It is critical that isolated males be kept isolated, and that community males be kept with their usual neighbors when these fish are not in use.

D Interpretation

Which hypothesis is supported by the results of the experiment? If we are interested in the behavior of the community males – that is, males who have territories adjacent to other males – why do we need to observe the behavior of isolated males? If we are interested in the behavior of community males toward novel intruders, why do we need to observe the behavior of community males placed back together after a short separation? Do you think the dear enemy phenomenon is an important factor in animal behavior? What special sensory and memory capabilities are necessary for the dear enemy principle to function?

III Male aggression and female mate choice

We can separate sexual selection into two types: selection for characters that contribute to success in male–male combat or competition, and selection for traits that females prefer (Karino, 1997; Akagawa et al., 1998; Doutrelant and McGregor, 2000; Doutrelant et al., 2001; Clotfelter et al., 2006). Examples of the former can include armaments such horns, antlers, and sheer strength. Examples of sexual selection by female choice include elaborate plumage (think peacock) or other showy physical features, and intricate behavioral displays.

Male mating success is often dependent on defeating competing males for the right to breed. This can take the form of vigorous display among males, with the female choosing her favorite, to all-out violent combat in which the victor claims the spoils.

In previous exercises, we have observed male–male competition in *Bettas*. Today, we will create a situation in which males must compete for the attention of a female by displaying. We will allow the female to demonstrate her preference for one of the males. Our main question is whether the more aggressive, vigorously behaving males are also the same males that females tend to prefer.

A Goals

- To observe male displays and female choice
- To create a hierarchy of male aggression and a hierarchy of female choice based on your experimental data
- To determine if these hierarchies are similar, i.e. are aggressive males also the most preferred males?

B Questions and hypotheses

Develop an explicit hypothesis and null hypothesis regarding the relationship between male/male aggression and female mate choice for this exercise.

C Methods

The male and female *Bettas* have been kept visually isolated. Each round of this experiment consists of two parts. First, pair two males for one minute and observe their behavior. After the time is over, decide which of the two males appeared to behave more aggressively throughout the observation period. Try not to assign a tie, but if the fish do appear to be evenly matched, this is acceptable. Visually isolate these two males and allow them to rest for a few moments. Next, place the two males on opposite ends of a female's tank (Figure 12.2). For one-minute record the amount of time the female spends on either half of her tank. After one minute, determine which male she preferred by noting in which half of her tank she spent the most time. Record the results of both trials in the matrices in Worksheet Table 12.2 and repeat with another pair of males.

After every possible combination of males has been completed, construct a hierarchy of male aggression and a hierarchy of female preference.

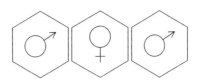

Figure 12.2 Schematic diagram of placement of hexagonal tanks for tests of mate choice in *Bettas*.

D Interpretation

Can you describe some examples of intersexual and intrasexual selection in other animals? The fish we use have been artificially selected for hyper-aggression for centuries and might fight to the death if we allowed them to do so. What do you suppose an encounter between males in the presence of a female in their natural habitat might be like?

IV Laboratory write-up

This is an excellent lab for a written report, and your instructor may ask you turn one in. All students should do all three experiments, but choose one to write-up. So that you'll have enough data for statistical analyses and your instructor may have all the students in the class pool their data for the write-ups.

References and suggested reading

Akagawa, I., Kanda, T., and Okiyama, M. (1998). Female mate choice through spawning parade formed by male-male competition in filefish *Rudarius ercodes*. *J. Ethology* **16**, 105–113.

Baird, T. A., Fox, S. F., and McCoy, J. K. (1997). Population differences in the roles of size and coloration in intra- and intersexual selection in the collared lizard, *Crotaphytus collaris*: influence of habitat and social organization. *Behav. Ecol.* **8**, 506–517.

Barber, I. and Wright, H. A. (2001). How strong are familiarity preferences in shoaling fish? *Anim. Behav.* **61**, 975–979.

Clotfelter, E. D., Curren, L. J., and Murphy, C. E. (2006). Mate choice and spawning success in the fighting fish *Betta splendens*: the importance of body size, display behavior and nest size. *J. Ethology* **112**, 1170–1178.

Doutrelant, C. and McGregor, P. K. (2000). Eavesdropping and mate choice in female fighting fish. *Behaviour* **137**, 1655–1669.

Doutrelant, C., McGregor, P. K., and Oliveira, R. F. (2001). The effect of an audience on intrasexual communication in male Siamese fighting fish, *Betta splendens*. *Behav. Ecol.* **12**, 283–286.

Forslund, P. (2000). Male–male competition and large size mating advantage in European earwigs, *Forficula auricularia*. *Anim. Behav.* **59**, 753–762.

Goldstein, R. J. (2001). *"Bettas."* Barron's Educational Series Inc., Hauppauge, NY.

Gomez-Laplaza, L. M. and Morgan, E. (2000). Laboratory studies of the effects of short-term isolation on aggressive behaviour in fish. *Mar. Freshwater Behav. Physiol.* **33**, 63–102.

Hojesjo, J., Johnsson, J. I., Petersson, E., and Jarvi, T. (1998). The importance of being familiar: individual recognition and social behavior in sea trout (*Salmo trutta*). *Behav. Ecol.* **9**, 445–451.

Hsu, Y. Y. and Wolf, L. L. (2001). The winner and loser effect: What fighting behaviours are influenced? *J. Anim. Behav.* **61**, 777–786.

Jaroensutasinee, M. and Jaroensutasinee, K. (2001). Bubble nest habitat characteristics of wild Siamese fighting fish. *J. Fish Biol.* **58**, 1311–1319.

Karino, K. (1997). Female mate preference for males having long and symmetric fins in the bower-holding cichlid *Cyathopharynx furcifer*. *Ethology* **103**, 883–892.

Leiser, J. K. and Itzkowitz, M. (1999). The benefits of dear enemy recognition in three-contender convict cichlid (*Cichlasoma nigrofasciatum*) contests. *Behaviour* **136**, 983–1003.

McGregor, P. K., Peake, T. M., and Lampe, H. M. (2001). Fighting fish *Betta splendens* extract relative information from apparent interactions: what happens when what you see is not what you get. *J. Anim. Behav.* **62**, 1059–1065.

McMillan, W. O., Weigt, L. A., and Palumbi, S. R. (1999). Color pattern evolution, assortative mating, and genetic differentiation in brightly colored butterflyfishes (Chaetodontidae). *Evolution* **53**, 247–260.

Peake, T. M., Matos, R. J., and McGregor, P. K. (2006). Effects of manipulated aggressive "interactions" on bystanding male fighting fish, *Betta splendens*. *Anim. Behav.* **72**, 1013–1020.

Switzer, P. V., Stamps, J. A., and Mangel, M. (2001). When should a territory resident attack? *Anim. Behav.* **62**, 749–759.

Utne-Palm, A. C. and Hart, P. J. B. (2000). The effects of familiarity on competitive interactions between threespined sticklebacks. *Oikos* **91**, 225–232.

Warner, R. R. and Dill, L. M. (2000). Courtship displays and coloration as indicators of safety rather than of male quality: the safety assurance hypothesis. *Behav. Ecol.* **11**, 444–451.

Warner, R. R. and Schultz, E. T. (1992). Sexual selection and male characteristics in the blueheaded wrasse, *Thalassoma bifasciatum* – mating site acquisition, mating site defense, and female choice. *Evolution* **46**, 1421–1442.

Wilson, S. (2002). Nutritional value of detritus and algae in blenny territories on the Great Barrier Reef. *J. Exper. Mar. Biol. Ecol.* **271**, 155–169.

"I'd make an awful soccer ball."

Worksheet 12.1 Territoriality.

Create a graphical representation of aggressive behavior for each of the three situations (visually isolated, male versus male, male versus mirror). How would you best represent the data? Choose the type of chart that you feel best allows you to interpret and explain your data. What should the X-axis and Y-axis show, and how should the number of aggressive acts be calculated so you can make comparisons between treatments?

Replicate	Aggressive acts by previously isolated males (number of acts)	Aggressive acts by community males versus novel intruders	Aggressive acts by community males versus their own neighbor
1			
2			
3			
4			
5			
6			
7			
8			
9			
10			

Worksheet 12.2 The dear enemy principle.

Versus	B	C	D	E	F	G	H	I	J
A									
B									
C									
D									
E									
F									
G									
H									
I									
J									

Versus	B	C	D	E	F	G	H	I	J
A									
B									
C									
D									
E									
F									
G									
H									
I									
J									

Have your TA show you how to analyze this data using an Analysis of Variance. What does the statistical result tell you about your data?

Which hypothesis does your data support? Why?

Worksheet 12.3 Female mate preference.

Male versus Male Aggression

Pair males and observe their interactions. Record the identity of the dominant male for every interaction. Do this with all possible combinations of males.

Male

Versus	B	C	D	E	F	G	H	I	J
A									
B	▨								
C	▨	▨							
D	▨	▨	▨						
E	▨	▨	▨	▨					
F	▨	▨	▨	▨	▨				
G	▨	▨	▨	▨	▨	▨			
H	▨	▨	▨	▨	▨	▨	▨		
I	▨	▨	▨	▨	▨	▨	▨	▨	
J	▨	▨	▨	▨	▨	▨	▨	▨	▨

Place a female so that she can choose between a pair of males. Do this repeatedly until she has chosen between all the possible combinations of males. In the table below, record the identity of the male in which the female is more interested

Versus	B	C	D	E	F	G	H	I	J
A									
B	▨								
C	▨	▨							
D	▨	▨	▨						
E	▨	▨	▨	▨					
F	▨	▨	▨	▨	▨				
G	▨	▨	▨	▨	▨	▨			
H	▨	▨	▨	▨	▨	▨	▨		
I	▨	▨	▨	▨	▨	▨	▨	▨	
J	▨	▨	▨	▨	▨	▨	▨	▨	▨

How do the hierarchies for male/male aggression and female choice compare? What does this result mean for the *Betta* breeding system?

Chapter 13
Learning

When an animal stores information for future use, it has learned. The ability to learn allows animals to cope with changing environmental conditions, and learning plays a key role in the ecology of many animal species. The study of learning is, perhaps, the biggest area of overlap between biology and psychology (Pearce, 1997). What distinguishes a biological approach to learning from a psychological approach? In biology, we would prefer to ask how learning functions in an animal's natural context, while psychological approaches to learning typically remove an animal from its context to allow precisely controlled experiments that focus on mechanisms of learning. In reality, these two approaches aren't as far apart as you might think; the mechanisms that can be identified in laboratory studies often accurately reflect the underpinnings of learned behavior by animals in the field.

Learning can take many forms, including extinction of a response to a stimulus through repeated stimulation (**habituation**), association of a new stimulus with an action (**classical conditioning**), and association of an act (or operation) with a reward or other response from the environment (**operant conditioning**). Operant conditioning can be viewed as a laboratory version of **trial and error learning**, which is often observed in the field. **Insight learning** is more difficult to measure, as it involves using a cognitive process to deduce the solution to a problem. In this lab we'll address each of these types of learning.

Habituation is a simple mechanism that animals can use to sort out an important stimulus from an unimportant stimulus. A stimulus that occurs repeatedly, but which turns out to have no particular significance, is soon ignored. Knowledge of habituation is important in understanding the behavior of wild animals which share environments with humans, as these animals are usually habituated to humans and don't show the response that they would normally exhibit to a nearby potential predator. For example, prairie dogs that live near roads or paths respond much less often with alarm calls than ones who live in more remote settings. We won't study habituation in this exercise, but it is important to keep habituation in mind as a mechanism that explains much in animal behavior. In training domestic animals, habituation often interferes with attempts to train dogs and cats, as they may habituate to the training regime and don't progress in learning the task.

Pavlov was one of the earliest scientists to study animal learning, and in all likelihood you've heard of his experiments on conditioned responses in dogs. These laid the groundwork for much of our understanding of animal learning. Pavlov presented hungry dogs with food, preceded by a bell. The dogs salivated at the sight of the food. Soon, the dogs began to salivate after hearing the bell. Pavlov's dogs' association of the ring of a bell with food, and salivating in response, is an example of **classical conditioning**. The bell is the conditional stimulus (CS), the food is the unconditional stimulus (US), the dog's natural reaction to the food, salivation, is the unconditional response (UR), while its learned reaction of salivating when hearing the bell is a conditioned response (CR).

Another important framework for understanding learning is **operant conditioning**, which differs in important ways from training using classical conditioning. Rather than just measuring a response to a stimulus, animals are trained to perform a task for a reward. By rewarding certain behaviors, experimenters can train animals to perform these behaviors. For example, pigeons can be trained to peck a key to receive a food reward. Mice, rats, or pigeons pressing levers in a box is an example of operant conditioning. In this lab, we'll explore conditioning using the ability of Siamese fighting fish, *Betta splendens*, to associate taps on their aquarium with food. To allow further exploration of conditioning, but avoid the complications of using living subjects, we'll turn to a computerized simulation, "Sniffy the Rat", as a way of testing how operant conditioning can be used to develop an understanding of learning mechanisms.

One interesting sidelight of this exploration will be to learn how operant conditioning can be used to "shape" Sniffy's behavior to do things that are, for a rat, unusual, such as turning a somersault. Clicker training, a commonly used technique in working with domestic animals, is also effective in "shaping". In this lab you'll use your labmates to explore clicker training as a technique for shaping behavior.

1 Goals

- To gain experience conducting learning experiments in animal behavior
- To test the principles of operant conditioning on *Betta splendens*, including the acquisition and extinction phases of learning

■ To learn the basic principles of the Skinner box and clicker training techniques in animal learning

II Learning in _Betta splendens_

In today's exercise, we will start an experiment with _Bettas_ to decide whether these fish are capable of learning by operant conditioning. There are several published studies of learning in _Bettas_ (Ducker et al., 1974; Bronstein, 1986a, b; Bronstein, 1988; Demarest, 1992).

We'll use _Bettas_ to ask questions like: How long does it take an animal to develop a conditioned response? And once it has developed, what happens when the reward is no longer given? These questions introduce us to the concepts of acquisition and extinction. Acquisition is simply the development of the conditioned response to the conditional stimulus. Acquisition might be affected by the duration of the training period, the frequency of the training period, the desirability of the reward, and, of course, differences among individual organisms. Extinction, in this context, refers to the cessation of learned behaviors after the behavior is no longer appropriate. For example, if Pavlov stopped feeding his dogs after ringing a bell, how long would it take for the dogs to stop salivating every time they hear a bell? If our fish can perform conditional learning, we will also investigate acquisition and extinction of the response.

A Methods

Record your results from these experiments on the Worksheet 13.1.

Operant conditioning – 1

The first phase of this experiment involves training the fish to feed from inside a white ring. As a lab group, decide how often you will perform the training exercise with your fish. Before you feed the fish, lightly tap the tank. Then, place one food pellet inside the ring. This can be repeated a number of times (the fish will usually eat up to six pellets). Follow this procedure every time you feed your fish. Observe the behavior of the fish each time you feed it. Does tapping the tank cause the fish to approach the ring? Record how the fish responds to a last tap each day on the worksheet. Remove the ring between feeding times/days.

If your fish learns to approach the ring after you tap the tank, you are ready to move to the next phase of the experiment. Continue reinforcing the fish's behavior by feeding it inside the ring for seven days. Then, present the fish with the ring, and tap on the tank, but do not feed it inside the ring. How long does the fish continue to go to the ring after the reward has been removed?

Operant conditioning – 2

Some students may wish to train the fish to do a different behavior than approaching a feeding ring. Another option is to train the fish to swim through a ring in order to receive a reward. Start by placing a large ring vertically inside the aquarium. Present a small reward, such as a bloodworm, on the opposite side of the ring, and allow the fish to swim through the ring to get the reward. After the fish performs the task regularly, slowly reduce the size of the ring. Finally, present the fish with the ring, but do not place the food on the opposite side of the ring. Rather, wait until the fish swims through the ring before rewarding it. Remove the ring when you are not actively rewarding the fish for swimming through it.

B Questions and hypotheses

Develop hypotheses regarding conditional learning in your *Betta*. Consider how the length of the acquisition period might vary with different reinforcement schedules. Further, once the fish has developed a conditional response, how does the length of time you reinforce the response affect the extinction of the response?

C Interpretation

The learning curve exhibited by the fish in this context is typical of trial and error learning. While the curve is best demonstrated by looking at the average of a number of animals, the variation among animals is interesting, as well. Compare your data with other students in the lab. Is a quick learner more "intelligent"? What might account for the variation among animals in their speed of learning? Animal behaviorists use conditionally trained animals in a variety of other experiments. Can you develop some questions and design experiments that you could do with your fish after they have been trained that make use of the fish's conditioned response to a stimulus?

III Sniffy the Rat

In addition to setting up your experiment on learning in *Bettas*, we'd like you to try a computer program that simulates learning by a rat in a Skinner box (Alloway et al., 2004). This program allows exploration of many complex aspects of both classical and operant conditioning. Given the time limitations for this laboratory, we'll explore the following aspects of learning:

■ Classical conditioning of a response (an association of sound and food).
■ Shaping an action by associating it with the previously learned sound.
■ Extinction of that action by removing the sound, the food, or both.

The study of learning in a Skinner box seems to be quite abstract – very much removed from real-life learning contexts. The processes you're learning, however, are just what you would use to train a pet dog or cat to do "tricks" or to be obedient, particularly if you're using a clicker in training your pet.

A Meeting *Sniffy the Virtual Rat*

The software you're using, and its associated manual, is called *Sniffy the Virtual Rat*. He's installed on the computers in the teaching lab; you can open the program by clicking on the rat icon in the dock (menu bar at the bottom or side of the screen). The computer may then ask your permission to change the resolution of the monitor; that's okay, so click to make that change. You'll then get a window of a Skinner box with Sniffy running around inside. Beside the window is a bar graph, which builds as you train the rat, first to the sound and then to push the lever. The bar graph gives you a measure of how associations between behavior and reward are building. Below is a strip chart window, which displays the speed of learning (the steeper the slope, the faster the learning) and cross-ticks that show when the food rewards are given to Sniffy. A third window, the lab assistant, gives tips about how to best train Sniffy. If these windows are not present, you can open them under the "Windows" pull-down menu. The "Cumulative record" and "Lab assistant" options are directly under this menu, you get the bar graph by going through "Mind windows" to "Operant associations".

Sniffy will be wandering around the Skinner box. There's a bar (which moves, clicks, and delivers a food pellet when pressed), a tray for food pellets, a light, a speaker, and a lever. If you press the space bar on your computer, you get the same click as when the bar is pressed, and food pellet is delivered. Try this out.

B Task 1 – Save your rat

Using the Save command, save your rat on the computer's desktop, using a name like "Mike's rat". No file extension is necessary.

C Task 2 – Learning to associate the sound with the food reward

Train Sniffy to associate the click with getting food. You can operate the space bar and record the total amount of time Sniffy takes to learn the task (he has completed learning when the bar graph for sound learning reaches the top of the graph). Tracking the amount of time from when you start training to when the red bar for sound training is at the top of the scale – will allow us to have a little contest to see who can train their rat the fastest. You should now be able to "call" your sound-trained rat to the food tray by clicking the space bar.

D Task 3 – Did you learn?

While Sniffy is learning, you're learning as well. Open a new window and save this new rat (different file name, of course). Train the second rat. How does the training time for your team compare with first rat? With other students?

E Task 4 – Shaping

Open your sound-trained rat and using "Save as" from the File menu, save a copy with the word "shaping" in the file name. You now need to take advantage of the fact that Sniffy occasionally rears up to teach him to press the lever, on his own, to obtain food. Go into the "Design operant conditioning experiment" menu and choose "bar press" as the reinforcement action. Each time Sniffy presses the bar he will get a reward.

Now you can compare whether it does any good to help him out by delivering rewards for rearing or being near the food with what happens if he's learning on his own.

Again, while you use the space bar to reinforce his rearing behavior (this is called shaping), you can record the total time, from beginning of training to when the bar-sound bar in the bar graph is all the way to the top. How hard is this? What strategies do you evolve to speed the training? Save the "shaped" Sniffy with a new file name using the "Saved as" option.

Now, open the "shaping" file. This time see how Sniffy does on his own, with no help from you in the shaping. You can speed up the process substantially by choosing the "Isolate Sniffy" option, but then the time comparison will be difficult.

You can use the various options under the "Reinforcement action" submenu in the "Design operant conditioning experiment" to teach Sniffy tricks, if you want.

F Task 5 – Extinction

Can these tasks be unlearned? For unlearning, or extinction, you have two options. The first is just to turn off the food supply. To do this, click the "Extinction" radio button in the "Design operant conditioning experiment" menu and then choose the "Mute pellet dispenser" option. What is his immediate response? How about his response over time, as his actions go unrewarded? Compare this with what happens when you punish (with an electrical shock) the bar press. You can leave the food on and punish the bar press, thereby giving contradictory feedback to him, or you can turn the food off and punish him. Try these different combinations.

G Summary and implications

If you explore Sniffy in detail, you'll see that once a behavior is learned, you can eliminate the reward (the food) and just use the conditioned stimulus (the sound) to maintain the learned behavior (Worksheet 13.2). You'll also see that intermittent or unpredictable reinforcement is very powerful in maintaining behavior; this is why slot machines are so attractive to some people.

◉ IV Clicker training

Animal trainers use "clicker training" to achieve the same results that you obtained with the "Sniffy" simulation. The animal is trained to associate the sound of a clicker with a food reward. The sound of the clicker is then used as a reinforcer for the desired behavior – sitting by a dog, for example. The food can be eliminated as a reward – the association between the click and the positive reinforcement is strong enough that the reinforcement doesn't actually have to be given. In clicker training, timing and consistency are everything. The click needs to come exactly when the animal is performing the desired behavior. Clicking at inappropriate moments, such as after the behavior, confuses the animal and greatly slows training. Negative reinforcements – a harsh word, for example – also confuse the animal and slow training. With care and patience, clicker training works well, though, and follows the principles we established with Sniffy.

A good website for instruction in clicker training is: http://clickertraining. com/training/clicker_basics/

Unlike training an animal with a clicker you can just specify that a click is a reward for a correct action. Remember, though, in animals that an association between reward and clicking must be built, and then that the animal will accept the click as a reward. You'll work with a partner in this part of the lab. The objective is to explore how clicker training works, trying to project yourself into the position of a dog, horse, or other animal that is being clicker trained. Your instructor will have a container with a list of possible physical actions, such as "rub the top of your head with your right hand". One student of the pair will randomly choose an action by drawing from the container, but will not disclose it to the other student. The first student will then attempt to clicker train the second student. Take turns at this, so that each student has tried to train the other student to do three or four actions. The students should agree in advance about how to record the data from these trials so that the class can then discuss how the clicker training worked.

A Discussion

As a class, you should consider questions such as: Are some students better clicker trainers than others? Better trainees? What personality characteristics make a good trainer or trainee? Can you extend this discussion to predict what kinds of animals might be effectively trained using this method?

References and suggested reading

Alloway, T., Wilson, G., and Graham, J. (2004). "Sniffy the Virtual Rat Pro, Version 2.0." Thomson, New York.

Bando, T. (1993). Discrimination of random-dot texture patterns in bluegill sunfish, *Lepomis macrochirus*. *J. Comp. Physiol. A* **172**, 663–669.

Bronstein, P. M. (1986a). Socially mediated learning in male *Betta splendens*. *J. Comp. Psychol.* **100**, 279–284.

Bronstein, P. M. (1986b). Socially mediated learning in male *Betta splendens* 2. Some failures. *Bull. Psychonomic Soc.* **24**, 306–308.

Bronstein, P. M. (1988). Socially mediated learning in male *Betta splendens* 3. Rapid acquisitions. *Aggressive Behav.* **14**, 415–424.

Demarest, J. (1992). Reassessment of socially mediated learning in Siamese fighting fish (*Betta splendens*). *J. Comp. Psychol.* **106**, 150–162.

Ducker, G., Nuttebau, N., and Schulze, I. (1974). Influence of aggression on learning and forgetting in *Betta splendens. Experientia* **30**, 747–749.

Pearce, J. M. (1997). "Animal Learning and Cognition: An Introduction," 2nd Ed. Psychology Press, Hove, East Sussex.

Worksheet 13.1 A learning curve for *Bettas*.

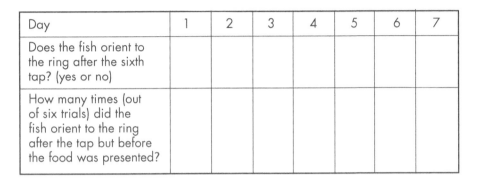

Day	1	2	3	4	5	6	7
Does the fish orient to the ring after the sixth tap? (yes or no)							
How many times (out of six trials) did the fish orient to the ring after the tap but before the food was presented?							

1. In the context of our experiment, what is the: conditional stimulus? Unconditional stimulus? Conditioned response? Unconditioned response? Why is it a good idea to remove the ring between feeding times?

2. How did varying the reinforcement time affect the length of the extinction of the response?

Below, draw two learning curves. The first should be based on the combined data for all the fish used by your lab section – take the percentage of "yes" answers for each day to construct the curve. The second should be based on your fish alone, taking the percentage out of the six trials for each day.

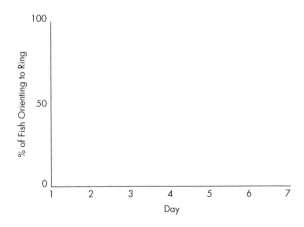

Worksheet 13.2 Sniffy.

Learning to associate the sound with the food reward
Do you think that the simulated behavior of Sniffy is an accurate reflection of how a real rat might behave in Skinner box? Why or why not?

Does this type of learning give us any insight into the behavior of animals in the wild?

Did you learn?
Well, did you? If your performance didn't improve, why?

Shaping
How does shaping Sniffy relate to teaching a pet tricks?

Extinction
Were you surprised by the rate at which extinction occurs?

Chapter 14
Analyzing sound and communication in crickets

Of all the communicatory modes, we can most easily analyze sound using reasonably high-tech methods (Hopp et al., 1997; Webb, 1998; Pollack, 2000; Gerhardt and Huber, 2002). In this lab we'll record sounds, represent them oscillographically, synthesize sounds using a signal generator, and use natural and synthesized sounds in behavioral tests for the ability of crickets to orient to calls of other crickets (Figure 14.1).

Animals produce sounds in many ways. Birds and mammals have cords inside their respiratory system that vibrate to produce vocalizations. Male

Figure 14.1 A female cricket. The tympanum, or ear, is the light colored spot on the leading edge of the front tibia. The ovipositor, which deposits the eggs, is between the cerci at the cricket's hind end. Males, of course, lack an ovipositor.

broad-tailed hummingbirds' wings strike together to make a distinctive high-pitched whir. Some snakes hiss by forcibly expelling air from their lungs. Cicadas sing using an external membrane that acts something like a drum. Grasshoppers and crickets chirp by rubbing body parts together. Termites make clicking noises by bashing their heads on the ground. All of these techniques produce vibrations that travel as waves in the air; the receiving animal's ears receive and interpret their signal value.

Sound has two important physical dimensions – frequency (sometimes called pitch) and amplitude (sometimes called volume). Using a microphone to receive sounds, frequency and amplitude can easily be translated into electrical signals. We'll display these signals using software that represents the sound waves in electrical form as a display on the computer screen. In this sonographic representation, time (frequency) is the x-axis and amplitude is the y-axis (Figure 14.2).

Cricket chirps are fairly simple communicatory signals (Bailey and Field, 2000; Shaw and Herlihy, 2000; Bateman, 2001). Male crickets are equipped with files – a series of ridges – on one wing and a scraper on the other. By moving their wings in the right way, they produce a single chirp. The waveform of the chirp tends to be characteristic of the species – we'll be using *Acheta domestica* – so that within the chirp there is a pattern of relatively constant frequency and amplitude (Figure 14.2). This waveform may contain information used by the receiving crickets to interpret the signal. Crickets normally produce their chirps in rapid series, so the time interval between chirps may also be quite important for the receiving animal. In our experiments, we'll be able to vary both within-chirp waveform and between-chirp time interval.

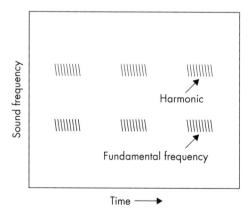

Figure 14.2   Sonogram of a series of three chirps in a cricket 'song'. By moving the scraper across the file, the cricket produces a series of clicks, which together make a single chirp. The frequency of the sound does not change – the cricket produces the same musical tone for all three chirps. Each chirp has a harmonic, a reflection of the main chirp at a higher sound frequency.

Because crickets are cold-blooded, the speed of their metabolic processes depends on the temperature of their surroundings. Cold crickets chirp more slowly than warm crickets. This creates an interesting problem in communicatory biology because the sensory expectation for the nature of the signal must change depending on the temperature at which the signal is produced. How do you suppose crickets accomplish this accommodation?

Male crickets produce chirps in territorial combat with other males and to attract females. We'll use cricket communication to explore phonotactic behavior. Phonotaxis is the ability to locate and move towards a sound source. This sounds like a simple task, but it is actually a fairly complex sensory and behavioral challenge. Sound, unlike visual signals, is for many animals not easily localizable to a point source. From your own experience, you know that sounds may echo and reverberate, causing confusion about their source. Unlike the eye, which contains many thousands of individual visual receptors, each responding to light from a particular location, each ear produces only a single sensory response at a given moment.

Animals locate the sources of sounds by comparing the intensity of sound between their ears. With ears on either side of their body, most animals can simultaneously compare the intensity of sound and determine the general direction of a sound source. This usually isn't adequate to precisely locate the sound, and most animals move either their head (if the ears are on their head) or their entire body to improve their ability to localize the sound by increasing the distance between their samples of the sound. Again, you know from personal experience that moving your head or moving around helps you to locate a sound source.

Crickets face the same problems in sound localization that we do, and perhaps it's even more difficult for them because their ears are so close together. One of the more interesting aspects of this lab will be observing the strategies used by crickets to locate and move towards a sound source.

▮ | Goals

- To gain familiarity with sound analysis and production software
- To explore the ability of crickets to discriminate between natural signals and synthetic signals, and to determine the critical parameters for crickets in identifying signals
- To analyze phonotactic behavior in crickets

II Questions and hypotheses

After gaining some experience with the software, examine the waveforms of recorded cricket signals that we'll provide to you. Then consider these questions:

- What might you vary in synthetic signals that would allow you to generate hypotheses about the crickets' ability to hear differences among sounds?
- Do male and female crickets respond differently?
- How will a cricket respond when there are two sound sources?
- Does it make a difference if the sounds are produced at different volumes?
- Does blocking one or both of the tympani with a little soft wax (provided in the lab) affect phonotactic behavior?
- Can you form your own hypotheses about the strategies that a cricket would use to find a sound source in the arena, and then test them?

III Methods

You'll be provided with a computer loaded with the appropriate software, microphones, speakers, a simple circuit board that will allow you to switch sound among speakers and to vary volume, crickets, and a 0.5 meter diameter arena in which the sounds may be played (Morris and Bell, 1982). Your assignment is to test the phonotactic abilities of five male and five female crickets (Worksheet 14.1). Then pick on of the above hypotheses or one of your own (be sure to check it with your instructor) to test (Worksheet 14.2).

IV Interpretation

As a class, discuss your results, starting with these questions: How do strategies for phonotaxis in crickets and humans compare? Can you propose some generalizations about how animals locate sources of sound? Put these thoughts into the context of a signaler who must call mates, but at the same time needs to avoid becoming food for a predator. How does predator avoidance affect the signaling strategy? Can you think of examples of predators that use sounds made by their prey while hunting? What adaptations do the predators display that might improve their phonotactic abilities?

References and suggested reading

Bailey, W. J. and Field, G. (2000). Acoustic satellite behaviour in the Australian bushcricket *Elephantodeta nobilis* (Phaneropterinae, Tettigoniidae, Orthoptera). *Anim. Behav.* **59**, 361–369.

Bateman, P. W. (2001). Changes in phonotactic behavior of a bushcricket with mating history. *J. Insect Behav.* **14**, 333–343.

Gerhardt, H. C. and Huber, F. (2002). "Acoustic Communication in Insects and Anurans: Common Problems and Diverse Solutions." University of Chicago Press, Chicago.

Hopp, S. L., Owren, M. J., and Evans, C. S. (1997). "Animal Acoustic Communication: Sound Analysis and Research Methods." Springer-Verlag, New York.

Morris, G. and Bell, P. D. (1982). Cricket phonotoxis: experimental analysis of an acoustic response. In "Insect Behavior, a Sourcebook of Laboratory and Field Exercises" (Matthews, J. R. and Matthews, R.W., Eds.). Westview Press, Boulder, Colorado, pp. 127–133.

Pollack, G. (2000). Who, what, where? Recognition and localization of acoustic signals by insects. *Curr. Opin. Neurobiol.* **10**, 763–767.

Shaw, K. L. and Herlihy, D. P. (2000). Acoustic preference functions and song variability in the Hawaiian cricket *Laupala cerasina*. *Proc. Roy. Soc. London, Series B, Biol. Sci.* **267**, 577–584.

Webb, B. (1998). Robots, crickets and ants: models of neural control of chemotaxis and phonotaxis. *Neural Networks* **11**, 1479–1496.

Worksheet 14.1 Phonotaxis.

Trace the paths of five male and five female crickets orienting to a vocalization coming from one speaker. Use different colored pencils to represent different crickets.

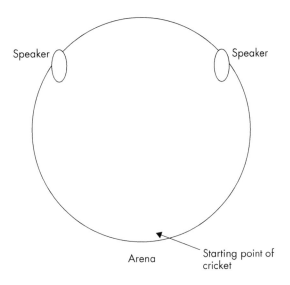

Do males and females differ in their orientation strategies?

Can you suggest how you would quantify the animal's paths for statistical analyses?

Lab, Worksheet 14.2
Phonotaxis.

Use this figure to illustrate your test of one of the other hypotheses concerning orientation behavior in crickets

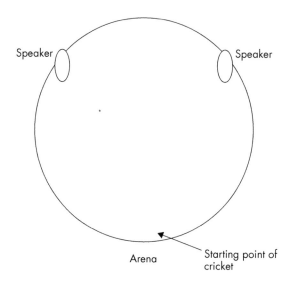

State your hypothesis, and your alternative hypothesis, in formal terms.

What do you conclude from your data?

Appendix: Notes on using Raven

Editing/manipulating a file

1. You can visualize the spectrogram in color by going to View > Color Scheme. Pretty cool.
2. To zoom in and out (so you can see the waveforms on the oscillogram) use the + and − buttons in the lower right corner of the sound window. These are at the end of the horizontal scroll bar.
   You can magnify the y-axis using the + and − buttons at the base of the vertical scroll bar.
3. The sound window can be resized (this is great for getting a larger view) by grabbing the bottom right corner and dragging. This is great for getting a better look at the oscillogram and spectrogram.
4. The easiest modification of the sound can be done by first selecting a segment (use the mouse and drag over a piece of the oscillogram). Then cut (command-V) or copy (command-C) the segment. This segment can be pasted into the sound at any point. If you copy a quiet interval, you can paste it repeatedly to increase the time between syllables. Cutting a quiet segment decreases the time between syllables. If you paste a quiet segment in the middle of a syllable, it will break that syllable with a pause.
5. You can change the recording's overall pitch by speeding up (makes sound higher pitch) or slowing down (lowers the pitch) the playback. This is done by changing the rate (default value 1.0) in the box at the upper right of the sound window.
6. You can filter a sound selectively to remove the selected frequencies (or all the frequencies except the selected). This allows you, for example, to remove either the fundamental or the harmonics. The menu path is: Edit > Filter > Around Active Selection or Edit > Filter > Out Active Selection.

Downloading/opening audio files

1. You can download audio files in various formats (wav, aif) and then open them in Raven. How this is done varies with the browser; in general, if you hold down the control button on the keyboard and the mouse button, a little window will pop up that allows you to save the linked file onto the desktop.
   There are lots of fun sound files at:
   http://www.grsites.com/sounds/animals001.shtml
   and
   http://www.freesoundfiles.tintagel.net/Audio/b-animals/
2. Move the file from the desktop to the documents folder.
3. Use the Open command in Raven to go through the file hierarchy until you find the documents folder, then open the file.

Recording sounds

1. Plug the iMic sound board into a USB port, and plug a microphone into the iMic.
2. Go to System preferences > Audio > Input, and select the Microphone as the input.
3. Open a recorder window in Raven, and hit the record button. Now any sound that comes into the microphone will be recorded. The recording can be saved and edited.

Chapter 15
Animal personality and intelligence

Recent scientific studies have verified what we knew all along; your pet (whether it's a dog, cat, snake, or turtle) has its own unique personality. You may also believe that your own pet is a pretty darn smart animal. Intelligence testing in animals is trickier than personality testing, and as you proceed with this lab, you'll find that personality and our perception of intelligence interact in interesting ways. Sih et al. (2004) argue for using the broader term "behavioral syndrome" to characterize sets of correlated behavioral patterns; we use "personality" because it has more intuitive meaning.

The idea that you can test an animal's personality and/or intelligence has contributed to the growth of applied animal behavior as a field, and practitioners commonly use personality tests to determine the suitability of an animal for a particular family, to screen candidate dogs for service roles, and to select animals for use in breeding programs. In this lab we'll explore aspects of personality and intelligence in domesticated animals.

I Goals

- To test for personality traits in dogs
- To design, implement, and report on an intelligence test for a family pet
- To gain a deeper understanding of what "personality" and "intelligence" mean in animals

▣ II Personality

Before using personality tests we should ask: What is personality? Can animals have personality? Personality is a set of attributes – such as sociability, aggressiveness, and willingness to please – which come together to form the social behavior of a species. What makes personality interesting is the variation of its expression among animals within a species, population, or social group. Scientists working on social behavior of birds or mammals are often struck by differences in personality among their study animals. This is particularly true of social animals such as primates, canids, felids, parrots and their relatives, crows and their relatives, and dolphins, but such variation can be found in a broad, and sometimes surprising, range of animals.

If personality varies among animals within a species, what function might this variation have? Variation may be the expression of different strategies, as predicted by game theory. Within this hypothesis, there are two possibilities. First, it may well be that success as a dominant animal calls for a different personality than does success as a subordinate, and expression of personality depends on status within the social group. In this type of system, an animal's personality may vary depending on the circumstances. Second, personality may be fixed genetically for a given animal, but it may vary among individuals because strategies differ in their success depending on environmental factors. If personality varies among animals, but is genetically fixed for an individual then the study of personality lies partly within the realm of behavioral genetics.

Among non-human animals, personality is best known in chimpanzees and domestic dogs. In chimpanzees, personality is best described by these variables (Weiss et al., 2000):

- Dominance
- Extraversion
- Dependability
- Emotional Stability
- Agreeableness
- Openness

The last five of these dimensions describe human personality (Bouchard, 1994); their presence in both chimps and humans probably reflects the shared evolutionary history of chimpanzees and humans. In both humans and chimpanzees, these personality traits have relatively high heritabilities and show virtually no effect of rearing environment. Human twins who are separated at birth and reared in very different environments show startling similarities in these aspects of personality. In their study of chimpanzees, Weiss et al. (2000) found a particularly strong heritability for social dominance and weak heritabilities for the other dimensions of chimp personality. As in human studies of personality, Weiss et al. (2000) found little effect of environment (in this case, different zoos) on personality.

In sum, personality in animals is real, measurable, and seems to be strongly influenced by genes. Variability in personality is, in a sense, genetic variability. This suggests that different personalities can be successful and persist in evolutionary time; if only one personality type were successful, natural selection would eliminate this variation.

III Methods: personality tests

Personality tests are well developed for dogs and primates; in this lab you'll use domestic dogs to learn how personality tests on dogs are conducted following techniques developed by Svartberg and Forkman (2002) and Svartberg et al. (2005). Once you've done the tests on a dog, you can ask how you might do similar tests on cats or other animals.

Your instructor will have obtained permission from your school's animal care and experimentation committee (often called the Institutional Animal Care and Use Committee or IACUC) to conduct these experiments, and the cooperation of an animal shelter to allow students in the class to test animals. The tests described here meet the ethical standards of the field; once again we stress that it is important as you conduct the tests to show respect for the animals and their welfare. Treat your companion animal with respect and, as you work out models for testing, make sure your tests do not inflict pain or discomfort on that animal. You will have puppies (the preferred test subjects) or adult dogs for testing. It is generally thought that a puppy must be at least seven weeks old to be mature enough to respond in these tests. Alternatively, you may be assigned the testing part of this laboratory as an exercise to do at home, employing a family pet or a friend's pet, during a break from school.

It is also very important to use caution with the animals while doing the tests. All dogs are capable of biting and of doing so unpredictably. Do not extend your hand towards an unfamiliar dog until that dog has shown a willingness to greet you and to play, and never put your face close to, or on the same level as, the dog's face. The common habit of greeting a dog by crouching down and inviting the dog to lick your face invites disaster with an unfamiliar dog. The instructor will demonstrate appropriate techniques.

Svartberg and Forkman (2002) identify five personality dimensions for dogs: playfulness, curiosity/fearlessness, chase-proneness, sociability, and aggression. They use 10 tests, each of which gives a number of scores, to place dogs along these personality dimensions. In this lab we'll use a simplified set of tests that will identify how four of these dimensions describe any given dog. These tests are listed in the Worksheet 15.1 for this lab.

IV Methods: intelligence tests

We often talk about how smart (or not) our pet animals are. Can you develop techniques to quantitatively test the intelligence of dogs or cats? This laboratory is designed as a homework exercise to be done with an animal that you know (such as a family dog, cat, or horse) and with which you have a bond of trust.

Intelligence testing in humans is a very controversial topic largely because it is quite unclear what an intelligence (or IQ) test actually measures. In humans intelligence testing is nested within language usage and human intelligence tests are very sensitive to differences among people in educational and cultural backgrounds. Tests can focus on very different aptitudes such as spatial reasoning, mathematical problem solving, reasoning with words (such as analogies), and the ability to plan and work through multi-step tasks. Thus there can be strong arguments that intelligence testing in humans is irrelevant, because of the lack of a consensus on what makes up intelligence, or that measures of intelligence are so inherently culturally biased that they do more harm than good. Nevertheless, it is probably clear to you from your experience that some people are quicker to solve problems than others or have high aptitude for certain activities like playing chess or music.

In animals, we can start a discussion of intelligence without worrying too much about the overlay of language and culture. A number of "intelligence" tests have been suggested for dogs and cats; you can view a typical set of dog intelligence tests at: http://abc.net.au/animals/dog_test/default.htm

This website includes descriptions of the tests and brief video demonstrations. The tests, as developed by Stanley Coren (2004) are:

1. The time which a dog takes to knock over a can that conceals a treat (the dog is allowed to observe the experimenter putting the treat under the can).
2. The time it takes for a dog to remove a bath towel thrown over its head.

3. The time it takes for a dog to respond to a human smile by wagging its tail and approaching the smiling person.
4. The time it takes for a dog to retrieve a treat from under a towel (as in test 1, the dog is allowed to watch the experimenter put the treat under the towel).
5. The time it takes for a dog to figure out that it can use its paw to collect a treat from under a low table.
6. The ability of a dog to respond to its own name, but not to an irrelevant word.

As you watch the videos on the website, one concern should be immediately apparent; most of these tests could be confounded by the dog's activity level and lack of timidity as much, or more, than intelligence. The eager, active dog will score well on most of the tests while the timid, shy, dog will do very poorly. Shyness does not reflect lack of intelligence, so you'll need to focus your thoughts on how to design intelligence tests that are not confounded by how outgoing or introverted your animal may be.

One of the most interesting observations from the videos on this website is how invested the humans are in the "intelligence" of their dog. They act in much the same way that you might imagine them reacting to a test of their child's intelligence. This level of human to dog empathy is an interesting part of the mix. As you design your intelligence test, make sure you don't give yourself the opportunity to bend the data by cheerleading for your animal to be "smart".

Another possible way to measure intelligence is by looking at speed of learning. In the lab on learning (Lab 14) we describe a process to construct a learning curve. A learning curve has time or the repetition number of a trial on the x-axis and the probability of success (displaying the desired behavior or solving the problem) on the y-axis. Through repeated trials, if a task is learned, the chances of displaying the appropriate behavior should go up. Can your dog quickly learn to do a new task on command, such as turning around, or does it take many repetitions? Use Worksheet 15.2 as a basis for your data collection.

Of course, if you've never trained your dog to do simple actions, such as shaking hands, sitting up, or lying down, it may be completely befuddled when you try this, not because it is dumb but because it hasn't learned how to learn in this way. So you need to train your animal in how to learn in response to rewards (operant conditioning), and once it has mastered the training routine you'll get a better measure of how many repetitions it takes to learn a new task. The website http://www.petrix.com/dogint/intelligence.html gives a ranking of dog breeds on the basis of how many repetitions of a task a typical animal in that breed takes to learn the task.

Another way of looking at this is to see if the animal obeys a command it knows the first time you give the command. A dog that is a little slow or is unmotivated to perform on uptake may require repeated commands to elicit the desired behavior. Of course, you should also consider that it may just be stubborn or hard of hearing.

V Interpretation

Develop your intelligence test and prepare a short presentation for the class using the following outline:

1. A clear statement of what measure of intelligence you use and why you think your test actually measures intelligence
2. A description of the design for the test
3. A summary of how your animal (dog, cat, horse, hamster, whatever) performed on the test
4. A comparison of your animal with other animals if you had others available for testing
5. A summary and conclusion about the effectiveness of your test.

In class, each student will present their test. Following the presentations, the class will discuss the following:

1. What is intelligence, anyway?
2. Which tests were most effective and why
3. How might you design an effective test of animal intelligence, based on the test designs presented in class.

Finally, if you want to see some examples of animals that are really good at learning, check out:

http://www.edge.org/3rd_culture/pepperberg03/pepperberg_index.html

http://www.metacafe.com/watch/28643/zac_the_athletic_parrot/

http://animal.discovery.com/fansites/petstar/videogallery/winners/video.html

References and suggested reading

Bouchard, T. J. (1994). Genes, environment, and personality. *Science* **264**, 1700–1701.

Coren, S. (2004). "How dogs think: Understanding the Canine Mind." Free Press, New York.

Sih, A., Bell, A. M., Johnson, J. C., and Ziemba, R. E. (2004). Behavioral syndromes: an integrative overview. *Q. Rev. Biol.* **79**, 241–277.

Svartberg, K. and Forkman, B. (2002). Personality traits in the domestic dog (*Canis familiaris*). *Appl. Anim. Behav. Sci.* **79**, 133–155.

Svartberg, K., Tapper, I., Temrin, H., Radesater, T., and Thorman, S. (2005). Consistency of personality traits in dogs. *Anim. Behav.* **69**, 283–291.

Weiss, A., King, J. E., and Figueredo, A. J. (2000). The heritability of personality factors in chimpanzees (*Pan troglodytes*). *Behav. Genet.* **30**, 213–221.

Wilsson, E. and Sundgren, P. E. (1997). The use of a behaviour test for selection of dogs for service and breeding 2. Heritability for tested parameters and effect of selection based on service dog characteristics. *Appl. Anim. Behav. Sci.* **54**, 235–241.

Worksheet 15.1

Simple tests for four personality parameters of dogs. We have not included tests for aggression, as these are potentially dangerous to conduct.

Personality Trait Being Tested	Behavioral Test	Score, from 1–5	Interpretation
Sociability	1. Greets stranger. 1 = no greeting, 5 = enthusiastic greeting with tail wagging and jumping		High scores for very sociable dogs
Sociability	2. Walks with stranger (on leash). 1 = refuses to walk, 5 = completely willing to walk with stranger.		High scores for very sociable dogs
Playfulness	3. Plays with rag, when offered by a stranger (unleashed). Owner and stranger throw rag back and forth, and then the rag is thrown about 10 meters (35 feet) from the stranger. 1 = no interest, 5 = follows rag		High scores for very playful dogs
Playfulness	4. Same as 3, scored for grabbing rag. 1 = never grabs rag, 5 = plays tug of war with the stranger or owner		High scores for very sociable dogs
Chase-proneness	5. Tie a furry pet toy at the end of a long rope. Release the dog from its leash and then pull the toy away from the dog in a zig-zag pattern. 1 = doesn't follow the toy, 5 = immediately runs after the toy at high speed		High scores for chase-prone dogs
Chase-proneness	6. Same as 5, scored for grabbing toy. 1 = never grabs the toy, 5 = grabs and holds toy for at least 3 seconds.		High scores for chase-prone dogs
Curiosity/ fearfulness	7. A hidden tester rattles a chain against a piece of sheet metal or a metal bucket while the owner is walking the dog; the noise should come from 1–2 meters away from the dog (perhaps from behind a tree on a walking trail). 1 = dog hesitates, 5 = dog runs more than 5 meters away		High scores for fearful dogs
Curiosity/ fearfulness	8. Same as 7, but after the noise, the owner approaches and touches the metal or bucket and then calls the dog. 1 = dog approaches without further encouragement, 5 = dog will not approach, even when encouraged		High scores for fearful dogs

Worksheet 15.2

This is the general format for a learning curve. For dogs, each time interval (x-axis) should be a training session, and the y-axis should be the percentage of correct responses during that training session. Dogs get bored with repetitive tasks, so 5–10 trials per session is appropriate.

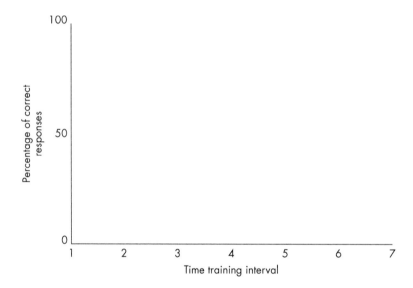

Chapter 16
Sensory physiology and behavior

Do all animals share essentially the same sense of taste? Or is the ability to taste tuned to a species' biology? In this laboratory we will explore the congruence of an animal's chemosensory perception with its feeding behavior. We will determine the sensitivity and selectivity of the sweet taste receptors in insects whose diets might include sweet-tasting resources such as floral nectar.

In general, animals are thought to have five basic types of taste receptors: sweet, sour, bitter, salty, and umami. Umami is the receptor for glutamate that gives food a meaty taste. Glutamate is also responsible for the flavoring effects of monosodium glutamate, which is sometimes used as a flavor enhancer in foods. Most of our knowledge of this range of flavor receptors comes from studies of mammals (Scott, 2004, 2005; Huang et al., 2006), including identification of the genes that code for the receptor molecules (Li et al., 2002; Kim et al., 2006). Pioneering work by Dethier (1963, 1976, 1987) on blowflies established a basis for broader comparisons of tasting abilities among animals and provided a foundation for subsequent research in fruitflies by Galindo and Smith (2001). However, much remains to be discovered about the sense of taste.

I Goals

- To learn lab techniques for detecting taste perception in insects
- To compare taste reception among species
- To place a species' tasting abilities within a larger ecological/evolutionary context

II Questions and hypotheses

We'll focus on the sweet receptor and once you've discovered where the taste receptors are located, we'll ask three questions.

- First, what is the concentration threshold for response to sucrose (table sugar) and how does this correspond to the concentration threshold for sweetness in humans?
- Second, do all sugars that taste sweet to humans elicit a response in your test animal?
- And, third, do artificial sweeteners such as sacharrin and nutrasweet (aspartame) taste sweet enough for your test animal to respond?

III The basic technique and model organisms for studying sensory physiology

In bees and some flies, the mouthparts form a straw-like structure called a proboscis used to drink liquid food. When the insect is not feeding, the proboscis is retracted into a groove on the underside of the head capsule. When food is perceived the proboscis is extended so that the bee or fly can drink. This simple response, called the proboscis extension reflex, gives investigators a powerful tool for studying chemical perception and learning in insects.

Three readily available insect species are ideal for this investigation: the honeybee (*Apis mellifera*), the housefly (*Musca domestica*), and the blowfly (*Phormia regina*). The fruitfly, *Drosophila melanogaster*, is also a possible experimental subject, but its small size and more fragile construction make it more difficult to use. Dethier (1963, 1976, 1987) used the proboscis extension reflex in his studies of blowfly sensory responses. More recently, the honeybee proboscis extension reflex has been used as a tool for studying learning (Gerber et al., 1996) and genetics (Rueppell et al., 2006). The housefly, while less studied, displays a similar response. Easily cultured butterflies, such as painted ladies, could also be used for these experiments.

The easiest way to manipulate any of these species is to glue a short stick to the dorsum (top) of the thorax. This provides a handle so that the insect can be held near a droplet of liquid and prevents your test animal from flying away. When the insect's taste receptors contact the liquid (it is left to you to determine where the taste receptors are located) it will respond by extending its proboscis if it perceives the liquid as having food value. You can collect your data as simple positive or negative responses.

Once the insect has extended its proboscis, it will start to drink the liquid. If you do repeated tests with the same insect, it will likely fill its crop (the insect term for stomach) and will gradually stop responding. This problem can be easily overcome by allowing your insect to initiate its proboscis extension reflex, but to remove the insect from the liquid food resource before it is able to consume any of the food. This will ensure that the animal does not become satiated. Alternatively, you may perform a minor operation to prevent the insect from sensing that its crop has become full, which will allow you to let your insect feed on the resource. Using dissecting scissors, make a small cut in between the first two abdominal segments and then puncturing the crop. This relieves the pressure inside the crop and results in an experimental subject that should respond consistently across a series of trials. It is probably best to perform this small surgery prior to any testing rather than wait until the fullness of the crop begins to inhibit the response.

IV Locating the taste receptors

Mammals taste by bringing potential food items into contact with their tongue. The sensory cells are arrayed on the tongue, concentrated on "taste buds", with different parts of the tongue being more sensitive to specific tastes. It would make sense for the insect system to be structured in an analogous manner, and careful observation will tell you whether the taste receptors are located on one of the more likely body parts – antennae, proboscis, front of the head, or the front legs.

V Concentration thresholds

Using table sugar, make a serial dilution so that you have the range of molar concentrations of sucrose. The molecular weight of sucrose is 342.3, so this quantity, in grams, dissolved in distilled water to make one liter of solution yields a one molar solution. One part of this solution in nine parts of distilled water gives a 10^{-1} molar solution, and so on. Using five test animals, start with the weakest (control) solution and test progressively stronger solutions, recording the data in the Worksheet 16.1. How does the threshold of your animals compare with your own threshold? Taking all of the data from bees tested by your class, graph the percentage of bees responding versus the concentration. Does the result suggest a step function, in which all bees begin to respond at the same concentration, or more of a linear relationship, which would suggest variation in thresholds among animals?

VI Range of responses to sugars

Make 1 molar solutions of the sugars listed in Worksheet 16.2. All of these sugars will taste sweet or semisweet to a human taster; how does your test animal respond? Using the combined data from your class, use a chi-square test to determine if there are significant differences in responses among sugars.

VII Artificial sweeteners

Artificial sweeteners taste sweet to humans, but perhaps not exactly "right". How do your test animals respond (Worksheet 16.3)?

VIII Further questions and points for discussion

1. Look online to see if you can find which sugars are found in nectars, and how do the responses of your test insect relate to the composition of nectar?

2. Do different sugars have different response thresholds? Using the sugars that yield responses in Worksheet 16.2, make serial dilutions and compare response thresholds.
3. If you have time, determine how the test animals respond to salty, bitter, umami, or sour tastes. Do any of these tastes elicit proboscis extension responses? Based on the differences between bee and blowfly diets, which one would you expect to respond to umami?
4. Following up on question 2, if you have time add another experimental animal, it would be interesting to compare response thresholds and the specificity of the sugar receptor between species.
5. Taking your collective results, what generalizations can you make about sweet receptors in animals?

IX Laboratory write-up

This is an excellent lab for a written report, and your instructor may ask you turn one in. Our suggestion is that all students do the first two experiments and then you choose one other experiment to perform. The write-up then should cover all three of your experiments and in addition to the usual coverage in your Discussion, you should address questions 1 and 5 in the previous section.

References and suggested reading

Dethier, V. G. (1963). "The Physiology of Insect Senses." Methuen & Co., London.

Dethier, V. G. (1976). "The Hungry Fly." Harvard University Press, Cambridge.

Dethier, V. G. (1987). Discriminative taste inhibitors affecting insects. *Chem. Senses* **12**(2), 251–263.

Galindo, K. and Smith, D. P. (2001). A large family of divergent *Drosophila* odorant-binding proteins expressed in gustatory and olfactory sensilla. *Genetics* **159**, 1059–1072.

Gerber, B., Geberzahn, N., Hellstern, F., Klein, J., Kowalksy, O., Wustenberg, D., and Menzel, R. (1996). Honey bees transfer olfactory memories established during flower visits to a proboscis extension paradigm in the laboratory. *Anim. Behav.* **52**, 1079–1085.

Huang, A. L., Chen, X. K., Hoon, M. A., Chandrashekar, J., Guo, W., Trankner, D., Ryba, N. J. P., and Zuker, C. S. (2006). The cells and logic for mammalian sour taste detection. *Nature* **442**, 934–938.

Kim, U. K., Wooding, S., Riaz, N., Jorde, L. B., and Drayna, D. (2006). Variation in the human TAS1R taste receptor genes. *Chem. Senses* **31**, 599–611.

Li, X. D., Staszewski, L., Xu, H., Durick, K., Zoller, M., and Adler, E. (2002). Human receptors for sweet and umami taste. *Proc. Natl. Acad. Sci. USA* **99**, 4692–4696.

Rueppell, O., Chandra, S. B. C., Pankiw, T., Fondrk, M. K., Beye, M., Hunt, G., and Page, R. E. (2006). The genetic architecture of sucrose responsiveness in the honeybee (*Apis mellifera* L.) *Genetics* **172**, 243–251.

Scott, K. (2004). The sweet and bitter of mammalian taste *Curr. Opin. Neurobiol.* **14**, 423–427.

Scott, K. (2005). Taste recognition: food for thought. *Neuron* **48**, 455–464.

Worksheet 16.1 Comparison of response thresholds to sucrose between test insects and humans.

Concentration of Sucrose Solution	Number of Insects (out of 5) Responding with Proboscis Extension	Number of Humans (Out of Whole Class) Sensing Sweet Taste
0 (distilled water control)		
10^{-6} molar		
10^{-5} molar		
10^{-4} molar		
10^{-3} molar		
10^{-2} molar		
10^{-1} molar		
1 molar		

Worksheet 16.2 Responses of test insects to a range of sugars*.

Sugar	Number of Test Insects (of 5) Extending Proboscis	Number of Humans (out of whole class) Sensing Sweet Taste	Qualitative Impression of Human Tester (e.g., is there an aftertaste or an undertone to the flavor?)
Five carbon sugars (Pentoses)			
Arabinose			
Xylose			
Ribose			
Six carbon sugars (Hexoses)			
Glucose			
Galactose			
Fructose			
Disaccharides (formed by combining two simple sugars)			
Sucrose			
Maltose			
Lactose			

*Mannose is particularly toxic to honeybees, but has some generalized toxicity and should be excluded from human taste tests. It should not be included in the experiment.

Worksheet 16.3 Responses of test animals to artificial sweeteners.

Sweetener	Number of Insects Responding (out of 5)	Qualitative Impression of Human Tester (e.g., is there an aftertaste or an undertone to the flavor?)
1 molar glucose (control)		
Nutrasweet (aspartame)		
Sacharrin		
Splenda (sucratose)		

Appendix I
Techniques in behavioral analysis: An overview

As you work through these labs, you will be introduced to many of the basic techniques used by scientists who work on animal behavior. They are summarized here as you may want to refer to the techniques as you develop your research project or use a technique discussed in one lab in another lab. You can also use this as a checklist to track your acquisition of skills through the semester.

■ 1 Observers and sampling

1. *Testing inter-observer reliability.* Commonly more than one person records data for a project. In the labs in this course, data from individual students or small groups is often pooled so that a larger sample size is available for analysis. If more than one person is collecting data for an experiment, are they all recording the same thing? You can try a simple experiment; use chalk to draw a line on a sidewalk where lots of people pass and then make a videotape of a 15 to 20 minutes of pedestrian traffic (or you can do this with cars at an intersection). Play the tape back to a group and have everyone separately count the number of people or cars crossing the line. Typically, the variation in the number is startlingly high. If you're doing a project that involves more than one observer, the key to reliable data is to train the observers well, cross-check their observations by occasionally

having more than one observer watch the same events, and compare data among observers to make sure you aren't getting inexplicably high variation among observers. **Never** have one observer watch only treatments and another watch only controls; this is a recipe for disaster.

2. *Blind observations.* A major issue in animal behavior studies is the subjective nature of how behavior might be interpreted. An important technique to eliminate observer bias in recording behavior is to have the observer be "blind" with respect to knowing the hypothesis for the experiment, whether the animal under observation is a treatment or control individual, and so on. For example, if you're comparing the dispersal behavior of recently fed sow bugs with the behavior of sow bugs that haven't eaten in a while, your experimental design would be better if you had a friend put each sow bug into a vial that is labeled with a letter or number. You could then release each sow bug and watch its movements, but you wouldn't know if you were watching a recently fed or hungry sow bug. At the end of the experiment, your friend would give you the code so you could identify which letters or numbers corresponded to recently fed or hungry sow bugs. Blind observations keep your hopes and expectations for the experiment from interfering with your experimental judgments.

3. *Random selection of experimental subjects.* In a legendary story, which may or may not have actually happened, a graduate student intent on studying activity in mice chose his experimental animals from a large cage which contained enough mice for the entire experiment. The first half of the group – the first ones he caught – he placed in his treatment group and the last half he placed in his control group. Sure enough, his treated animals were far less active than his control animals. What's wrong with this picture? His result may have been explained by the fact that less active animals were easier to catch. There's no way of knowing, based on the way he captured his animals, if his treatment actually had an effect on activity. To insure that catchability or other characteristics of animals does not affect results, it is best to assign a number to each potential experimental subject and then use a random numbers table, or the random function in Excel to determine the choice of animals for treatment and control groups. Your approach to randomization may vary depending on your species, how many animals are available, and whether you can capture or mark your animals; in any case, take every caution not to embed a bias in your experiment before you begin by improperly assigning animals to treatment and control groups.

4. *Pseudoreplication.* For an observation to be a true replicate, it needs to be independent (unaffected) by any of the other replicates. If you measure a behavior over and over again on the same animal, you don't generate a new replicate with each measurement, as the samples aren't independent. When this problem is present, the data are said to be "pseudoreplicated".

Under certain circumstances, repeated measures on an individual are allowed. This happens when your experimental objective is to see if a treatment affects each individual in your sample. These experiments focus on before-treatment and after-treatment measurements. For example, if you take a group of humans with high blood pressure, give them a medication that you think might help and then measure their blood pressures again, you have used a repeated measures design. There are statistical tests (the simplest is a paired-comparisons *t*-test) that are specifically designed to test hypotheses when you have repeated measures.

In animal behavior, studying social groups presents special problems in pseudoreplication. An issue that crops up repeatedly is that an investigator wants to study the dynamics of behavior in a flock, herd, school or colony of social insects. Inevitably, more than one measure of behavior is taken on each individual as observations are made. In these cases, the best approach is to have multiple social groups and to compare between groups, as well as within groups. Even though your within-group observations are pseudoreplicated, if all the groups exhibit the same patterns then the results are usually viewed as convincing.

5. *Sampling techniques*:
- *Scan sampling*. You visually scan across a group, recording what each animal is doing (Altmann, 1974; Altmann and Altmann, 1977). This is good for constructing time budgets and for measuring variation among animals. This method has the advantage of sampling, without bias, all the animals in the group (Chapter 7).
- *Ad lib samples*. This involves just writing down events as you see them (Altmann, 1974; Altmann and Altmann, 1977). The problem with this approach is that you may bias your sample to particularly obvious or unusual animals/behaviors.
- *Focal animal sampling*. You choose an animal and watch its behavior for a predetermined period of time (Altmann, 1974; Altmann and Altmann, 1977). This allows you to follow the behavioral sequence of an individual and to measure the length of time it spends in a location or performing an activity. The obvious disadvantage is that you're watching only one animal at a time, so your sample size is limited. You also can run into trouble with bias if you don't choose your focal animal randomly. You can overcome these difficulties by using focal animal observations on a number of randomly chosen animals (Chapter 7).
- *Instantaneous samples*. This is when the activity of all the animals present is recorded at a single time. If there are a large number of animals, either multiple observers or a photograph that can later be analyzed are good ways of obtaining instantaneous counts. Instantaneous samples were rarely done prior to easy digital photography, but now that photographs can be taken cheaply and analyzed on a computer screen, this method has become more prominent (Chapter 3). A series of photographs taken over time, for example, can be used to measure behavioral constancy of animals in a flock, school, or herd.

II Ethograms, time budgets, and sociograms

6. *Ethogram construction.* An ethogram is a catalog of the individual behaviors that an animal displays. To make an ethogram, watch the animal species in question behave, and as you see each new behavior, write down a descriptive name and a definition of the behavior. Your goal is to have good enough definitions that someone, working from your definitions, would be able to identify the behaviors by the correct name. Depending on your goals, ethograms can be more or less detailed. For example, you might watch a cat and define the following behaviors: **licks own paw**, **licks fur on own back**, and **rubs paws over own face**. This would be good if your goal is to compare the details of grooming activity among cats. If, on the other hand, you're interested in how cats allocate time among major activities like feeding and grooming, then you may wish to lump these acts into a single category, **self grooming**. Transition diagrams, sociograms, and time budgets are all extensions of the ethogram approach and are discussed below.

7. *Time budget analysis.* Once you have the ethogram, it is a simple matter to sit with a stopwatch and determine how much time an animal spends engaged in each behavior. Time budgets are very useful for asking if there are differences in activity among animals in a social group (due to division of labor or slacking by certain individuals) and for making seasonal comparisons. For example, do the time budgets change once nestlings hatch? Are there differences in male and female time budgets while they're caring for nestlings? Time budgets are a surprisingly powerful tool for gaining understanding about the factors that drive the behavior of animals.

8. *Transition diagrams.* In a transition diagram, you consider the likelihood that one particular behavior may follow another behavior. Usually the likelihood is expressed as a probability. For example, in a hypothetical species, a partial transition diagram among its major activities might look like Figure AI.1.

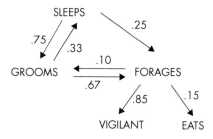

Figure AI.1   A sample transition diagram.

9. *Sociogram construction.* Like an ethogram, but a sociogram captures the behavioral interchange between a pair of animals. You start by making a good ethogram for the species. The sociogram is then often constructed by making a matrix of behavioral interactions between pairs of animals. In this example, you would use the rows to represent one animal in an interaction (Animal A) and the columns to represent the other animal (Animal B). Thus if Animal A approaches Animal B and B responds by approaching, you count this in the cell with an*.If B responded by fleeing, you would mark it in the cell with a # (Table AI.1).

By watching many interactions, you can fill in the table and then if you want, make a transition diagram based on the sociogram as illustrated below. Probabilities would then be assigned to each arrow in the transition diagram based on the matrix (Figure AI.2).

10. *Dominance hierarchy analysis* (Chapter 12). In dominance hierarchy analysis, a matrix of winners and losers of combat is used to determine dominance ranks. Below is a hypothetical dominance matrix in which Animal A is the dominant, beating B five times, C three times, and so on. D won no encounters. In this example the dominance hierarchy is linear, with A dominating all the animals, B dominating all but A, and so on. In field conditions, dominance hierarchies may not be so linear. Landau's *H* is a

Table AI.1 Format for constructing a sociogram.

		Behavior of Animal B			
		Approaches	Flees	Bares teeth	Bites
Behavior of Animal A	Approaches	*	#		
	Flees				
	Bares teeth				
	Bites				

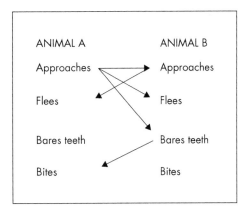

Figure AI.2 A sample interaction diagram.

Table AI.2   Using the results of aggressive encounters to construct a dominance hieararchy.

Winning Animal	Losing Animal			
	A	B	C	D
A	XXX	5	3	1
B	0	XXX	4	1
C	0	1	XXX	2
D	0	0	0	XXX

statistic that is sometimes used to measure the linearity of a dominance hierarchy and can be calculated using this formula:

$$H = 12/(n^3 - n) \sum (v_a - (n - 1)/2)^2$$

where N is the number of animals and V_a is the number of animals dominated by animal a. H varies from 0 to 1, with 1 reflecting a perfectly linear hierarchy, in which the top animal always dominates all others, the second always dominates all but the top animal, and so on. 0 reflects no hierarchy; any animal may dominate any other animal in the group. Table AI.2 illustrates how this is done.

III Behavioral ecology

11. *Measuring population size (Lincoln index)*. It is surprising how often animal behaviorists need an estimate of the size of a population. For example, if you're observing 10 animals, is this the entire population in your study area or only a small fraction of the population? The simplest technique for population estimates is called the Lincoln index, which relies on capture-mark-release-recapture technique. This index assumes that animals are neither entering nor leaving your population; if you think that the populations you're studying is open to immigration or emigration you'll need more complicated techniques that are described by Southwood (1978).

To use the Lincoln index, capture animals in the population (the number captured is N_1), mark the animals, and then release them. After enough time to allow for the marked animals to be randomly mixed into the population, capture a second sample (the number captured in the second group is N_2). M is the number of marked animals in the second sample. The estimate of the population = $N_1 \times N_2/M$. If you don't recapture any marked animals then you need to increase the number of marked animals, N_2, or both.

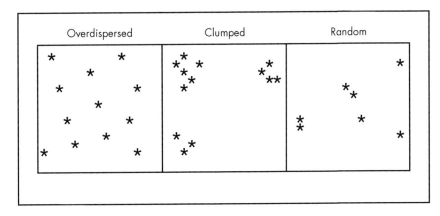

Figure AI.3   Overdispersed, clumped, and random distributions.

Capturing and marking animals will likely require state and federal permission if they're birds or mammals; if you can recognize distinctive markings on the coat or plumage of individuals, then you can apply the Lincoln index technique doing the capture and marking procedures. This is best done by taking a photograph of a subset of the population (this gives you N_1) and later taking a second photograph. How many animals in the first photograph are still represented in the second photograph? You can use the Lincoln index on this data.

12. *Spatial distribution mapping.* Maps of spatial distribution are highly useful in helping to determine if animals are clumped, dispersed, or randomly distributed. This then can be used to form hypotheses about territoriality and about patterns of resource use. For animals that make nests or dens, it is often easier to map these physical structures than the actual locations of animals. Idealized distributions appear as in Figure AI.3.

Overdispersed distributions imply, but don't establish, that territorial interactions are occurring. Clumped distributions suggest clustering in favored habitats. Be careful though because scaling can affect your perception and animal may be territorial within a clump. Random distributions suggest that the presence of one animal does not affect the presence or absence of others.

The hypothesis that a spatial distribution is random can be tested by using nearest neighbor analysis. The techniques are not overly complicated and the details can be found in papers by Meagher and Burdick (1980), Sinclair (1985), and Dixon (1994).

Usage polygons are another approach that is valuable when you're interested in the relationships between territories and home ranges. This requires watching an animal and recording its location on a map at fixed time intervals. This is more feasible with rodents, such as chipmunks, prairie dogs or marmots that are visible during daylight hours and don't roam widely than with birds, nocturnal animals, or animals with very large home ranges. Once you have lots of points on the map, you can draw a polygon that encloses all the points; this is the 100% usage polygon and if you've collected enough data it is the animal's home range. Then starting from the center of the 100% polygon or the nest or den and moving out, count points until you reach half the total points. This is the 50% usage polygon. You can repeat this process to generate as fine a gradation of usage as you need. If you repeat this process with neighboring animals and then overlay the maps, the larger map will show you if home ranges overlap or not. [Borger et al. (2006) give a detailed discussion of sophisticated applications of this technique.]

IV Other useful tools

13. *Construction of a learning curve.* Learning curves are very useful for measuring how animals respond to experience or for making comparisons within and between species in learning abilities. The concept is simple; measure the success of an animal at a task over repeated trials or use a group of animals and measure the proportion of the group that succeeds. Graphing the results allows you to ask whether the animal's responses to a stimulus change with experience. Graphs like this are found in Chapter 13 and take the general form given below, although the time interval can be changed (seconds, minutes, days, etc) (Figure AI.4).

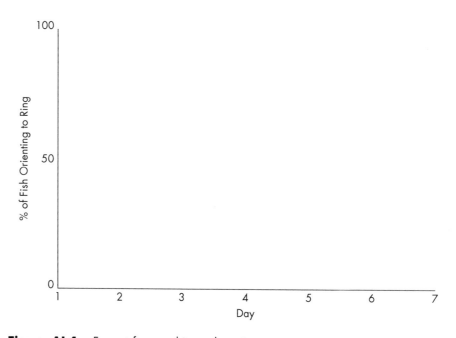

Figure AI.4   Format for graphing a learning curve.

14. *Recording and playing back vocal signals.* This is a powerful technique for testing communicatory value of animal signals. Do animals respond to playbacks? If so, in what contexts. You can use playbacks to test interesting hypotheses concerning orientation and communication. The Raven software (Chapter 14) allows you to manipulate vocalizations; does changing the gap between syllables or phrases, for example, change the responses of animals to a vocalization?

15. *Manipulation of chemical signals (pheromones).* Pheromones are sometimes surprisingly easily manipulated in the field. Do you wonder if ants are using a chemical trail for orientation? Gently rub or sweep the surface of their pathway; often this is adequate to disrupt their chemical trail if they're using one. Do you wonder if dog urine is used in communication? Watch the behavior of dogs approaching a tree where one has urinated, then wash the tree trunk with water. Does their behavior change? Of if you live in an area where you get snow in the winter, use a shovel to move patches of "yellow snow" (Bekoff, 2001). Many insect pheromones have been chemically identified and can be ordered from chemical supply houses or pest control suppliers for use in investigations of pheromone-related behavior.

References and suggested reading

Altmann, J. (1974). Observational study of behavior – sampling methods. *Behaviour* **49**, 227–267.

Altmann, S. A. and Altmann, J. (1977). Analysis of rates of behavior. *Anim. Behav.* **25**, 364–372.

Bekoff, M. 2001. Observations of scent-marking and discriminating self from others by a domestic dog (*Canis familiaris*): tales of displaced yellow snow. *Behav. Proc.* **55**, 75–79.

Borger, L., Franconi, N., De Michele, G., Gantz, A., Meschi, F., Manica, A., Lovar,i S. and Coulson, T. (2006). Effects of sampling regime on the mean and variance of home range size estimates. *J. Anim. Ecol.* **75**, 1493–1405.

Dixon, P. (1994). Testing spatial segregation using a nearest-neighbor contingency table. *Ecology* **75**, 1940–1948.

Kearns, C. A . and Inoue, D. W. (1993). "Techniques for Pollination Biologists." University Press of Colorado, Boulder, Colorado.

Meagher, T. R. and Burdick, D. S. (1980). The use of nearest neighbor frequency analyses in studies of association. *Ecology* **61**, 1253–1255.

Sinclair, D. F. (1985). On tests of spatial randomness using mean nearest neighbor distance. *Ecology* **66**, 1084–1085.

Southwood, T. R. E. (1978). "Ecological methods : with particular reference to the study of insect populations," 2nd Ed. Wiley, New York.

Appendix II
Statistics

What kind of data might you collect? The first, and simplest type, is category data. This is a count of the number of occurrences of a thing, an event, or anything else that you can count. You might, for example, compare the number of males and females in a classroom. The categories are the genders and you obtain your data by counting how many of each is present.

Most other data are obtained by measurement. Anything quantified by a measurement scale, such as length, mass, volume, elapsed time, etc. yields this type of data. Two common descriptive statistics used to characterize measurement data are mean and variance. We characterize measurement data using two descriptive statistics, mean and variance. The mean, or average, of a series of numbers is simply the sum of the numbers divided by the number of observations. The amount of variation around the mean is described using the variance. The standard deviation and the standard error of the mean are calculated from the variance; variance, standard deviation, and standard error are mathematically related

---

Box AII.1 Variation around a mean.

s = standard deviation = (sum of squared deviations from the mean*)/sample size
s^2 = variance = square of the standard deviation
SE = standard error = standard deviation/square root of the sample size

*To calculate the squared deviation from the mean, you first calculate the mean of the sample. Then, for each replicate, subtract the value of the replicate from the mean. The result of the subtraction is the deviation from the mean. Then square each of these and add them together; this number is the sum of squared deviations from the mean for the sample.

and in animal behavior, scientists most often give the standard error to describe the variance around their averages.

The standard deviation tells us how much variation there is around the mean. The standard error takes into account your sample size and gives you an estimate of the variability that you would find in means if you did the same experiment repeatedly. Standard errors tend to be smaller if sample sizes are larger, an important point to remember as you design experiments.

I Statistical tests

The following statistical tests are particularly useful when you're learning how to study animal behavior:

- Chi-square – use when comparing how often events occur or to compare observed with expected frequencies
- t-Test – use to compare two averages
- Analysis of variance – compares three or more averages
- Correlation – tests for association between two variables
- Regression – also tests for association; in this case you control one of the variables

II Adding probabilities to hypothesis testing

Is just comparing counts or averages enough to actually test your hypothesis? The answer is clearly no. When you test a hypothesis, you want to know the likelihood that your conclusion is correct. Biological systems vary and a completely unequivocal result is rare, particularly in studies of animal behavior. We use statistical probabilities to help us to understand the comparisons we make in experiments.

Statistics allow us to determine the probability that our results might be due to chance. When you read about a biological experiment, usually you'll see an expression like "$p < 0.05$" given with the numerical result of the statistical test. Biologists usually consider results with $p < 0.05$ to be "statistically significant", meaning that we consider the probability of this result occurring by chance is less than 5%.

A key factor in making statistical comparisons is having an adequate sample size to test your hypothesis. In a laboratory course, obtaining the needed sample size (10 to start, often you'll need 20 to 30 for each thing you're measuring) is often difficult because of time limitations. One way around this is to pool together the results from several students; this is the approach we'll use in our labs. If you're doing a semester-long independent project, you'll want to talk with your instructor about how to set goals for sample sizes.

III Data analysis in Excel

We will use an example in which we are observing sparrows at a bird feeder. Our hypothesis is that the larger individuals will be dominant at the feeder. Note that this is a chicken-or-egg situation: If the hypothesis is supported, we don't know whether the larger birds are dominant because they're large or whether they are large because they are dominant and therefore able to get more food. In this case, a correlation analysis is appropriate because a correlation makes no statements about causality; it merely tells you whether there is a relationship or not.

A Entering data into Excel

Our data consists of the weights in grams of our test birds, time spent at the bird feeder over a 2-hour observation period (presence), and time spent actually eating at the feeder in that same period. We enter it into Excel so that each column is one of the above variables and each row represents one bird (Table AII.1).

Table AII.1 Sample data for summary statistics.

Weight (g)	Presence (min)	Feeding (min)
25	35	22
42	50	20
14	10	10
18	6	5
31	24	21

B   Summary statistics

To get a first quick look at our data, we'll do a **Sort** (under the Data) tab. We'll sort by weight; this will put our birds in order by weight and we can see whether our other variables also seem to increase as weight increases. Be sure to click "My list has a Header row" so that it doesn't include your labels in the Sort.

Usually, you start out looking at some basic properties of your data. The most common of these is the mean. To calculate the **mean**, or average, you would add all of the numbers together and then divide this by the number of subjects. For example, the mean feeding time is $(22+20+10+5+21)/5 = 15.6$.

In this case, there are few enough data points that you could calculate it by hand, but Excel will also do it for you. It will also provide you with some other numbers that give you a sense of what your dataset looks like. Go to Tools > Data Analysis* > **descriptive statistics**. Use the mouse to highlight your input range, the Feeding time column. If you include the column label in your highlight, be sure and check the Labels in First Row box. Check the Summary Statistics box, and specify an output range. Here's what it looks like for Feeding time (Table AII.2):

Table AII.2   Sample summary statistics output.

Feeding (min)	
Mean	15.6
Standard Error	3.414674216
Median	20
Mode	#N/A
Standard Deviation	7.635443668
Sample Variance	58.3
Kurtosis	−1.927776421
Skewness	−0.806700959
Range	17
Minimum	5
Maximum	22
Sum	78
Count	5

*If you don't have the Data Analysis option under Tools, go to Add-ins (also under Tools) where you can ask the computer to add-in the Analysis Tool Pak.

The output will give you three measures of **central tendency** – the **mean**, **median,** and **mode**. The mean is as described above; the median is the middle number in your data set if the numbers were all written out in order from smallest to largest, and the mode is the most commonly occurring number in your data set.

You will also get several measures of **Spread** or the amount of variability in your data. A very basic, but useful measure of spread is the range, which is simply the difference between the smallest and largest number in your data set. It will also give you the **variance** and **standard deviation**. These two are closely related and they basically express the average difference between all the data points and the mean. The smaller the variance or standard deviation, the better job the mean does of telling you about the data. The larger these numbers are, the more your data vary from the mean. A final measure of spread is the **standard error**, which is also related to the standard deviation but it takes into account your sample size. Generally, the larger your sample size, the smaller your standard error. In labs, you'll usually have some guidance on which of these numbers to report in a given situation. If you're working on an independent project, you should check with your instructor about which to use.

C Graphing your data

- Producing a graph of your data is easy using the Chart Wizard and it's a good way to get a quick idea of whether there's a relationship between your variables. With this type of data, a scatter plot is the best type of graph to use.
- Click the Chart Wizard icon (looks like a small bar graph) > *XY Scatter* > Next
- Using the mouse, select all data in the Weight and Presence columns > Next
- Enter the title you want to give the graph and the axis labels (including units of measurement). In this window, you can also play around with how the graph will look, for example, you can tell it not to show the legend (the legend is meaningless in this example) and to dispense with gridlines > Next > Finish.

Figure AII.1 shows what your resulting graph might look like.

A quick glance at this graph tells us that there may indeed be a relationship between the size of the bird and the time it spends at the bird feeder. In order to be able to say so with any confidence, however, we need to do some statistics. When you graph your results, Excel gives you

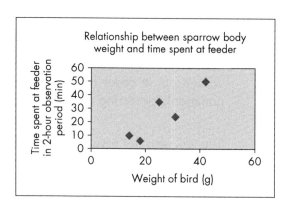

Figure AII.1   Sample graphical representation of a correlation.

the option of generating a **regression line and equation**; the regression is calculated to give you the line that best fits your data. You should use regression when you have controlled one of your variables and correlation when you have controlled neither of your variables.

D   Correlation analysis

When we're doing a scientific study, we collect data that we hope will tell us about relationships between variables. But the data we collect is always a partial picture. We cannot observe and measure every single sparrow at every single feeder in existence. We can only measure a sample that we hope reflects the true relationship between sparrow body size and feeder dominance. In this case, we only have five birds in our sample. How confident can we be that the pattern we see in our sample of five birds is the true pattern and not the product of random chance?

A **correlation** tells us whether there is a relationship between the variables *in our sample*. With a correlation, the outcome is a value between −1 and 1. The closer to −1 the value is, the more negatively correlated the two variables are (that is, there is a strong negative relationship). The closer to 1 the value is, the more positively correlated they are. If the value is close to zero, there is no strong relationship.

Tools > Data Analysis > Correlation

Use the mouse to select all the values in the Weight and Presence columns. If you include the column labels, be sure to click the Labels in First Row box. > OK

The output you get will be a correlation matrix. The number you're interested is in the second row, first column: 0.89424.

Since 0.89 is pretty close to 1, we can feel much more confident that there is indeed some relationship between our variables. As we saw in the graph, Weight and Presence appear to have a fairly strong, positive relationship, based on our sample. Larger birds spend more time at the bird feeder. Or, birds that spend more time at the bird feeder are larger.

Can you think of ways this study could be improved?

What other questions could we ask of these data that might help us get a clearer picture of what's going on at the bird feeder? Why did we measure Presence and Feeding times as separate pieces of data? What could the relationship between these two variables tell us?

E *t*-Test

In behavioral studies, we often want to compare two treatments. In this case, one of our variables is a categorical variable, that is, it is in one of two possible states, A or B. In the correlation above, both variables were continuous, meaning that their values could take on any numerical value. Let's do an example of a categorical variable.

We'll use the same sparrows as above, but this time we want to know how their feeding behavior changes if we use a different type of bird-seed in the feeder. (This is a rather mundane question, but it illustrates the point.) Using the same five birds, we are still measuring time at feeder, but on one occasion we set out sunflower seeds and on another occasion we set out millet (Table AII.3).

Table AII.3 Sample data for *t*-test.

Time at feeder (sunflower) in min	Time at feeder (millet) in min
4	15
12	15
17	18
19	22
24	29

A *t*-test basically compares the two averages of the two different treatments and tells us whether they are significantly different.

– Go to Tools > Data Analysis > *t*-test: Paired two-sample for means.

Once again, use the mouse to select your two input ranges, which in this case is the two columns. Check the Labels box if you included the column labels in your input ranges, and select an output location. Here's what the output looks like (Table AII.4):

Table AII.4   Sample output for a t-test.

t-Test: Paired Two Sample for Means		
	(sunflower) in min	(millet) in min
Mean	15.2	19.8
Variance	57.7	34.7
Observations	5	5
Pearson Correlation	0.867119757	
Hypothesized Mean Difference	0	
df	4	
t Stat	−2.673695691	
P(T < = t) one-tail	0.027795895	
t Critical one-tail	2.131846782	
P(T < = t) two-tail	0.055591791	
t Critical two-tail	2.776445105	

Excel gives you the mean for each of your treatments. Note that the mean for millet is 19.8 and the mean for sunflower is 15.2. So, on average, the birds spent longer at the feeder when it had millet in it than when it had sunflower seeds. But is the difference between these two averages "real", or could it have happened by chance? If we did the study again, could sunflower have the larger mean? The t-test gives us a sense of how unusual, and therefore meaningful, our results are. We need to look at two numbers from the output: the *t-statistic* and the *p-value*. The larger the t-statistic the more significant the difference between our means. You can disregard the negative sign on the t-statistic by the way, and just look at its value, 2.67. This is a fairly small value, as t-statistics go. The p-value tells us how likely it is that we could get the result we got by chance alone. A small p-value means that the probability of our result happening by chance is very small, and therefore there is probably a meaningful difference between our two means. In other words, the smaller your p-value, the more likely it is that you have detected a real pattern or effect. In general, the scientific community accepts a p-value of < 0.05 as "significant". One last confusing thing about a t-test: the output gives one-tailed or two-tailed p-values. You'll notice that in our output, the one-tailed p-value is significant, but the two-tailed falls just short of the 0.05 requirement. Which of these you get to go by depends on the structure of your experiment and your data; you will need to consult your instructor to determine which applies to your situation. In this case, you would probably report that there does appear to be a significant difference in the time spent at the feeder depending

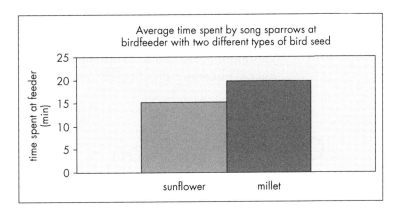

Figure AII.2   Graphical comparison of means analyzed by a *t*-test.

on the type of birdseed, but because of the small sample size, further investigation is warranted.

An analysis that uses a *t*-test is usually graphed as a bar graph with two bars (Figure AII. 2).

F Chi-square tests

The chi-square test works when you want to compare category, rather than measurement data. For example, you might divide a group of dogs into three categories – good learners, modest learners, and non-learners. After training the dogs in a way that you think will improve their learning abilities, you might then reclassify the dogs. Your data could look like Table AII.5.

Table AII. 5   Sample data for a chi-square test.

	Learning Ability Before Training	Learning Ability After Training
Good learners	5	13
Modest learners	12	6
Poor learners	3	1

We recommend using a website:

http://www.georgetown.edu/faculty/ballc/webtools/web_chi.html

rather than Excel for calculating chi-square tests. The website allows you to easily enter your data and calculate the chi-square. For this data, the website first asks you to enter the number of rows (3) and columns (2) of data. You then get a table where you can fill in the actual counts. Clicking a button results in the chi-square calculation; in this case, the chi-square value is 6.55, and the probability is less than 0.05; the hypothetical treatment significantly improved the ability of the dogs to learn.

G   Interpretation

OK, so you have done your statistics and you are ready to report your results. But what do your results mean? What have you learned from your study? In our hypothetical sparrow study, we have learned that song sparrows appear, on average, to spend longer at birdfeeders that have millet in them than at birdfeeders that have sunflower seeds. One might be tempted to interpret this as indicating that they prefer millet. However, one could also argue that sunflower seeds are a much larger and richer food source than millet. Therefore birds spend less time at the feeder simply because they're getting nourishment more efficiently. A third interpretation could be that the larger sunflower seeds require more handling and digesting, so the birds may come and get a seed or two and then go away and digest for awhile, whereas the smaller millet can be processed as they eat. So here we've done our study, but we still have all these questions! What we should learn from this example is: (a) one study often generates lots more hypotheses to test; (b) what we learn from an experiment depends heavily on the design of the experiment, that is, the results of your experiment should actually give you an answer to your question; and (c) think carefully about what your results mean and how else they might be interpreted!

Appendix III
A short essay on the use of vertebrates in teaching labs

Many students of biology are concerned about animal welfare and the use of animals in scientific experimentation. While much can be learned about animal behavior only by actually observing animals, the learning value of experiments must be weighed against animal welfare considerations. The biggest educational value from using living animals in teaching laboratories comes from the experience gained with the variability and unpredictability of living systems. Students who have this experience are well equipped to address scientific problems as their career progresses, whether they enter work in a science-related business, graduate school, or medical school.

The greatest concerns have been expressed about "higher" vertebrates such as birds and mammals. Because of student concerns and the logistical difficulties inherent in proper animal maintenance, we have opted not to design labs that employ captive birds or mammals. We feel strongly that the same principles and experiences can be developed while working with animals that do not have such obvious reactions that some people consider "pain" or "fear".

In many ways the learning objectives of this course could be accomplished using single-celled organisms, such as bacteria or protozoa. Our experience is, however, that many students are not satisfied with labs that only use invertebrates (even complex ones such as insects) as study subjects. Thus we have tried to balance student concerns with animal welfare, on the one hand, with student wishes to work with certain types of animals. We have addressed the student desire to have experience with vertebrate behavior by using Siamese fighting fish, *Betta splendens*, for several of the labs. We hope that rather than tiring of your fish, you grow attached to them.

Index

Printed and bound by CPI Group (UK) Ltd, Croydon, CR0 4YY

03/10/2024

01040311-0017